SOVIET
SECRET PROJECTS

FIGHTERS SINCE 1945

SOVIET
SECRET PROJECTS
FIGHTERS SINCE 1945

TONY BUTTLER & YEFIM GORDON

MIDLAND
An imprint of
Ian Allan Publishing

Soviet Secret Projects – Fighters Since 1945
© Anthony Leonard Buttler & Yefim Gordon, 2005

ISBN (10) 1 85780 221 7
ISBN (13) 978 1 85780 221 4

First published in 2005 by
Midland Publishing
4 Watling Drive, Hinckley, LE10 3EY, England
Tel: 01455 254 490   Fax: 01455 254 495

Midland Publishing is an imprint of
Ian Allan Publishing Ltd

Worldwide distribution (except North America):
Midland Counties Publications
4 Watling Drive, Hinckley, LE10 3EY, England
Tel: 01455 233 747   Fax: 01455 233 737
E-mail: midlandbooks@compuserve.com
www.midlandcountiessuperstore.com

North America trade distribution by:
Specialty Press Publishers & Wholesalers Inc.
39966 Grand Avenue
North Branch, MN 55056, USA
Tel: 651 277 1400   Fax: 651 277 1203
Toll free telephone: 800 895 4585
www.specialtypress.com

Design and concept
© 2005 Midland Publishing and
Stephen Thompson Associates
Layout by Sue Bushell.

Printed by Ian Allan Printing Ltd
Riverdene Business Park, Molesey Road
Hersham, Surrey, KT12 4RG, England.

Visit the Ian Allan Publishing website at:
www.ianallanpublishing.com

Photograph on half-title page:
The Ye-152A displays its twin side-by-side engine arrangement during its flyover for the Tushino display of 9th July 1961.

Photograph on title page:
The never-to-be-flown Yakovlev Yak-140 prototype.

Contents

Introduction
and Acknowledgements

Since 1990, following the break-up of the former Soviet Union, a great deal of new information has been made available that describes the Warsaw Pact's post-war military aircraft. This includes drawings of unbuilt fighter projects and a number of Russian-language publications have printed a good selection of these. Quite a few English-language books and journals have also reproduced drawings and material relating to certain designs, but it is believed that this is the first work to be published in English that gives full coverage to the post-war Soviet fighter projects which are currently available. There are still many gaps to be filled and, on occasion, different original sources have given conflicting information, particularly in regard to dates.

Coupled with these projects, and presented in a logical sequence, is the progress that has been made in the design and development of the Soviet and Russian fighter during the sixty years that have elapsed since the end of the Second World War. To set the scene brief reference is also made to pre-war and wartime Soviet work on jet engines and jet-powered aircraft. The book follows the style of earlier volumes in this series covering British and Soviet aircraft design. However, as noted on several occasions in the text, the Soviet Union turned a high percentage of its fighter proposals into prototypes which means that, to keep within the guidelines, this volume contains more accounts of real aircraft than previously (the cost of building so many different prototypes must have been colossal). This may bring accusations of rehashing but in fact a general overview of Soviet jet fighter design has not been produced for quite a while and here it benefits from previously unpublished material. The book also offers a chance to highlight some of the forgotten aircraft from the Soviet Union's history, like several of the Lavochkin prototypes. In general, Soviet military aircraft were never given names (there are a few exceptions) but for identification purposes the Western powers eventually introduced a codename system for known types and these are included for completeness.

At the time of writing the archives for the Mikoyan, Sukhoi and Yakovlev Design Bureaux must hold quite a lot more material on unbuilt designs that has not yet been made available. For some of the other projects it is a little difficult to decide just how important they might have been, but Nigel Eastaway considers that a project which reached the mock-up stage was a serious programme, and if any metal was cut it became very serious. This seems to be a very good definition to work to. Those readers interested in knowing more about the fighters that have served the Soviet Union's and Russia's armed forces are recommended to move on to the series of books published on individual types by Yefim Gordon, many of which are listed in the bibliography. As ever, it has been a very enjoyable experience to write about these superb aeroplanes, a good many of which represented the cutting edge of Soviet technology.

Acknowledgements

Unless stated, all of the illustrations in this book come from the Yefim Gordon collection. Special thanks must be given to Nigel Eastaway of the Russian Aviation Research Trust for his considerable help in finding extra information and filling gaps, to George Cox for allowing us to take photographs of the models in his collection (some of which were created by Alex Panchenko), and to Joe Cherrie and John Hall for kindly offering to make models that filled gaps in the illustration coverage. Sincere thanks also to Vladimir Rigmant (Tupolev archivist), Jens Baganz for his drawings, Peter Kostelnik for translations, Paul Martell-Mead, Thomas Mueller, Helmut Walther, Dr Peter Wolf and the authors of numerous articles published in *The Bulletin of the Russian Aviation Research Group*. We are also grateful to Brunolf Baade, his son Olaf Baade and Dr Reinhard Mueller (as publishers of the book *Dessauer Flugzeugprojekte 1945/46 fur die Sowjetunion*) for permission to use material relating to the Baade EF-series of projects. Finally thanks once again to the team at Midland for their support and to Keith Woodcock for his cover painting.

Tony Buttler, Bretforton
and Yefim Gordon, Moscow
September 2005

Chapter One

The Straight Wing Era

Although the subtitle of this book is *Fighters since 1945*, the story of the Soviet Union's turbojet engines, and therefore its first jet-powered aircraft designs, actually begins much earlier in 1937 – in fact the first Soviet proposal for a jet engine was made by B S Stechkin in 1929. To help understand what has happened since 1945, it is worth making a brief review of what took place during those preceding years.

The first detailed Soviet jet engine design was drawn by Arkhip M Lyul'ka at the Kharkov Aviation Institute (KhAI) in Ukraine and was called the RTD-1 (rocket turbojet engine); it used a centrifugal compressor and was intended to give 4.9kN (1,100 lb) of static thrust. At the same time the KhAI-2 fighter project was designed around this power unit and was intended to reach a top speed of 900km/h (559mph). A P Yeremenko, a KhAI undergraduate, worked with Lyul'ka on the fighter's layout and, although the Institute's scientific council was unimpressed, Lyul'ka was recommended to present the project to

officials in Moscow. There Professor V V Uvarov, a specialist who was heading research into turboprop engines, assessed the papers and rated the project highly. However, at this time insufficient resources were available to make the fighter a reality but the KhAI-2 project can claim to be one of the world's first jet-powered aircraft designs.

In 1938 Lyul'ka moved to SKB-1 (Special Design Bureau No 1), which was based at the Kirov Factory in Leningrad (now St Petersburg), and during the following year work began on the 5.1kN (1,155 lb) thrust RD-1 engine which used an axial compressor (this engine should not be confused with a later liquid-propellant rocket engine that received the same designation). Progress was swift and by May 1941 around 70% of the components for the first RD-1 had been manufactured, the combustion chamber and turbine were undergoing bench testing and construction of the compressor was under way. Unfortunately, Germany's invasion of Russia on 22nd

Above: **The second Yak-25 seen in flight with undertip drop tanks.**

Below: **Engine designer Arkhip Lyul'ka.**

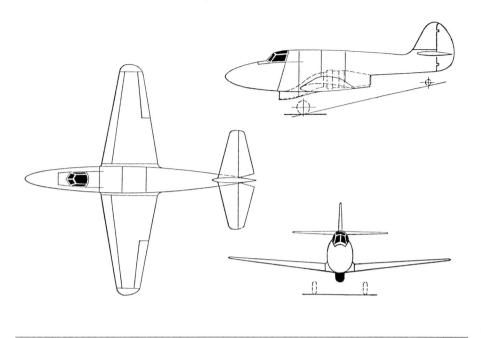

The KhAI-2 fighter project of 1937. This date ensures that the KhAI-2 can claim to be one of the world's first designs for a jet-powered aircraft.

Gudkov Gu-VRD (4.43).

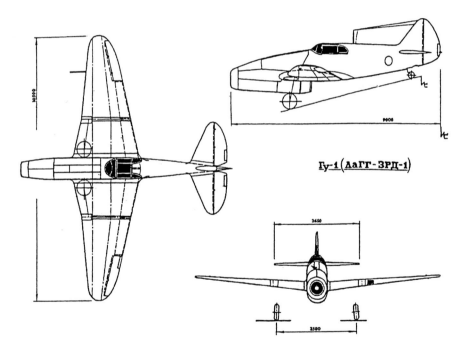

Гу-1 (ЛаГГ-ЗРД-1)

Model of the Lavochkin La-VRD. Joe Cherrie

June halted work, primarily because this new power unit was thought to be of little value. As German troops approached Leningrad, most of the Bureau's specialists were evacuated further inland and those engine components already manufactured were preserved and hidden away inside the Leningrad factory, where they remained until 1944.

This cost the Soviet Union precious time in the race to develop a jet. However, by late February 1942 Air Forces Command was insisting that work should resume on jet engine development and in August Lyul'ka's team returned to their work, this time at factory No 293 in Sverdlovsk (which has now been renamed Ekaterinburg). Here they proposed a 'turbo-compressor' jet engine that was expected to give 14.7kN (3,305 lb) thrust for a weight of 700kg (1,543 lb).

Gudkov Gu-VRD

In 1943 Mikhail I Gudkov began to design the Gu-VRD fighter, which was to be powered by a single example of Lyul'ka's latest engine mounted in a 'stepped' fuselage. Like many early jet designs all over the world, this was really a piston fighter airframe fitted with a jet instead of the piston unit, but Gudkov had previously worked with S A Lavochkin on the LaGG-3 piston fighter and knew what he was doing. The Gu-VRD had its engine placed in the bottom of the fuselage aft of the nose section, giving in side view a 'step' behind the engine nozzle after which the rearward fuselage had a much smaller cross section; as we shall see this arrangement was to be used by other first-generation Soviet jet fighters. The Gu-VRD was a single-seater and had a nose air intake and a tailwheel undercarriage and was armed with one 20mm ShVAK cannon and one 12.7mm (0.5in) BS machine gun; internal fuel load was 400kg (882 lb), estimated range 700km (435 miles) and the time to 5,000m (16,404ft) was 1.39 minutes.

On 10th March the project documentation was submitted to official circles but, after appraisal by the Council of People's Commissars, it was rejected, simply because it was felt that a suitable engine which gave the required power would not actually be available for some time. I I Safronov, one of the officials, wrote on 17th April 'apparently the aircraft would fly with the claimed speed but

Semyon Alekseyev played a modest role during the early years of Soviet jet fighter design.

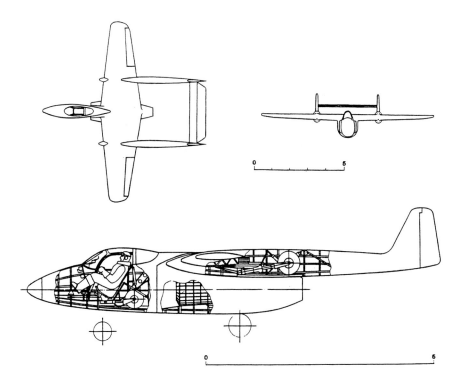

Lavochkin La-VRD (2.44).

the problem is that, as of today, there is no engine, just the name of its designer'. Thus, due to the delays with the Lyul'ka engine the Gu-VRD was never built, and by the time work on Lyul'ka's later S-18 power unit (below) was under way, Gudkov's OKB had been disbanded. VRD stands for Vozdooshno-Reaktivnyy Dvigatel, or Air Reaction Engine. Gudkov's presentation also noted that he was working on a fast bomber powered by two of Lyul'ka's 14.7kN (3,305 lb) thrust engines. This had an estimated take-off weight of 6,500kg (14,330 lb), a top speed of at least 780km/h (485mph) at 600m (1,969ft) altitude and a range of at least 1,200km (746 miles).

–

In the spring of 1943 Lyul'ka and his team moved to Moscow and on 20th May specialists from the People's Commissariat evaluated his new engine and judged that it was not sufficiently developed, so there was clearly no likelihood either of a prototype power unit being built. As a result, and to help with the overall development of this new form of propulsion, the Central Institute of Aviation Motors (TslAM) established a jet engine research laboratory during the following August with Lyul'ka as its chief.

Lavochkin La-VRD

Despite the fact that a complete jet engine still did not exist anywhere in Russia, the country's aircraft designers were by now showing great interest in this form of propulsion. In February 1944 the Lavochkin Experimental Design Bureau's Moscow Division at Aircraft Factory No 81 completed a drawing of a fighter project

called the La-VRD. This had a twin-boom layout (with an all-metal wing) and was one of the few jet aircraft designs drawn worldwide to employ this feature, although two of its contemporaries, the de Havilland DH.100 Vampire and SAAB J-21R, did in fact enter service. No twin-boom jet fighter design produced by an American company has ever been turned into hardware.

The Lavochkin project was designed from the start as a jet aircraft and appears to have been an attractive layout. The design effort was led by Semyon M Alekseyev, Lavochkin's closest aide, who very much liked the twin-boom arrangement having used it on an earlier series of piston fighter studies. However, it appears that V R Yefryemov did most of the work of creating the project. The La-VRD also incorporated TsAGI recommendations and data relating to the design of high-speed aircraft, information on which the design team was reliant in establishing the overall general arrangement; it also had a pressurised cockpit. The engine chosen for the La-VRD, the Lyul'ka S-18 described shortly, was fed by lateral air intakes and used a tricycle undercarriage with the main wheels retracting rearwards into the booms. One 23mm cannon was placed in the front of each boom and the engine was housed in the short main fuselage with the tail placed above the level of the jet efflux. A total of 500kg (1,102 lb) of fuel was to be carried. La-VRD was expected to reach 5,000m (16,404ft) in 2.5 minutes, have a ser-

vice ceiling of 15,000m (49,212ft) and range, when flying at top speed and sea level, of 140km (87 miles). It was the end of the year before the first bench test example of the S-18 was ready, which meant that the La-VRD project went on hold. In early 1945 Lavochkin began work on new fighter designs.

–

In the meantime the Soviet Union had learnt that the Germans actually had jet aircraft in front-line service, while information was also acquired about the progress that had been made on British and American jet programmes. In January 1944 a German staff car driver, Franz Warnbrunn, was captured by the Soviet army and revealed that he was in fact an engineer who had worked at the Heinkel aircraft factory. He was able to report that German jet engines and aircraft were already undergoing flight-testing. In addition, in April 1944 an American YP-59 jet fighter was demonstrated to two Soviet engineers, Kochetkov and Souproon, at the Bell aircraft factory in America. In truth, Russia still had no native turbojet or jet-powered aircraft, so the country's leaders took steps to fill this gap and try to catch up with its allies and enemies. In February 1944 the GKO (State Defence Committee) established the NII-1 Scientific Research Institute which brought together groups of engineers who had been working on different forms of 'jet' engine, turbojets, ramjets, pulsejets and liquid and solid-propellant rockets, and Lyul'ka was given control of the Institute's turbojet department.

Above: **Mikhail Guryevich and Artyom Mikoyan, the two great Soviet fighter designers pictured together.**

Left: **Semyon Lavochkin.**

The first step was to transfer all of the RD-1 components and documents from Leningrad to NII-1, but all they revealed was that the engine, in comparison to foreign designs, would be short on thrust and performance. On 22nd May 1944 the GKO issued a decree which stipulated that Lyul'ka had to produce a turbojet engine rated at 12.2kN (2,755 lb) thrust by 1st March 1945. Lyul'ka's response was the S-18 turbojet completed at the beginning of 1945. It featured an eight-stage axial compressor and the 'flyable' version was later designated TR-1, work on that being carried out during the early months of 1946 with the first bench run taking place in July. Tests revealed that the TR-1 gave a maximum thrust of 12.7kN (2,865 lb). During the spring of 1947 the follow-on TR-2 turbojet was designed. This offered a nominal thrust of 22.0kN (4,960 lb) and a maximum 24.5kN (5,510 lb). Four units were built but in the following year this engine's development was cancelled because series production had now begun on the RD-45 (the Soviet copy of the British Nene).

The end of the war altered the situation considerably. German jet aircraft and jet engines, and indeed complete factories specialising in jets, fell into the hands of the Soviet Army and from March 1945 the Special Central Directorate of People's Commissariat of Aviation Industry began combing Germany for equipment and machinery to send back to the USSR. This included specific searches for jet aircraft and engines and their documents and, later that summer, the State Defence Committee issued a decree which called for an evaluation of the results of German scientific research into jets; it also demanded further testing of captured jet aircraft and the copying and production of the BMW 003A and

Junkers Jumo 004 power units. By August 1945 NII-1 and TsIAM were bench testing German engines and by the end of the year quite a number of units had been brought to the Soviet Union. The Jumo 004 was also copied as the RD-10, under the leadership of V Ya Klimov, and the BMW 003A as the RD-20, and several upgraded versions and more advanced developments of both types were eventually produced by the Soviet Union.

Nevertheless, despite the progress that had been made, it was clear that the quality of the German engines fell below the world standards of that time and their application was generally considered to be a temporary measure. On 6th April 1946 the Soviet leader, Joseph Stalin, chaired a meeting at the Kremlin to discuss the status of jet aviation development in the Soviet Union. Native designs lagged behind developments overseas and, to improve the situation in the short-term, it was decided to try and obtain the most advanced jets available. Great Britain seemed to be the best partner in this respect since it was the world leader in turbojet development at the time. Since the end of the war relations between Moscow and London had cooled but the countries were still allies.

Soviet intelligence reported that the Rolls-Royce Derwent 5 and Nene 1 jet engines, the types in which Soviet experts were showing most interest, had entered full production and were being installed in the Gloster Meteor and de Havilland Vampire fighters (in fact Soviet intelligence was partly in error, the Nene was only ever fitted in a few Vampires and the standard engine was the de Havilland Goblin). These power units showed several advantages over German engines – longer life, higher thrust, and the like – and it was agreed that an attempt should be made to try

and acquire some examples. In due course Rolls-Royce agreed to supply ten Nenes and ten Derwents and follow-on orders took the eventual totals despatched to 25 Nenes and 30 Derwents. These were also copied as the RD-45 and RD-500 respectively, and were put to good use in a variety of new fighter and bomber designs. The Soviets also had a go at trying to buy a small number of Meteor and Vampire fighters, which were then considered to be the best of their type in the world. On 14th January 1947 the USSR Council of Ministers issued a decree to purchase three examples of each and the Soviet Trade Mission in London began negotiations with Gloster and de Havilland, but in this case an export licence was never granted.

Following these events, the next Soviet-designed jet engine was the Klimov VK-1, the development of which began in the middle of 1946, using the British Nene as a pattern. However, at the time the Klimov Design Bureau possessed very limited information on the Nene, essentially just articles and descriptions published by Western magazines plus some photographs which included a longitudinal cross-section. Moreover the specification requested a thrust of 26.4kN (5,950 lb): that is, about 20% higher than a Nene. After the first British engines had arrived in the USSR, the VK-1's design was modified and the first example was ready for its factory testing in October 1947, before the RD-45. In due course the VK-1 and its developments were built in massive numbers and supplied power to a whole generation of Soviet aircraft.

Before moving on to the post-war design of Soviet fighters, it is worth taking a quick look at the background to the fighter specialist OKBs, and a little of their relationship with the Government and its official establishments. In 1938 Semyon Lavochkin teamed up with V P Gordonov and M I Gudkov to produce the LaGG-1 and LaGG-3 piston fighters. These were followed by the very successful La-5 but by the end of the war Lavochkin's two partners had moved on. His Design Bureau then produced a long series of jet-powered fighter prototypes but it was never to recapture the glory of its wartime piston types. Lavochkin died in 1960 and the OKB carrying his name was switched to the production of spacecraft and missiles.

In 1939 Artyom Mikoyan was asked to form a new Experimental Design Department (or OKO) operating within the Polikarpov Design Bureau. The job he was required to do was to lead a team that was to design a new fighter but he was only prepared to accept if his close associate, Mikhail Guryevich, could join him.

He could and the MiG-1 (MiG standing for Mikoyan and Guryevich) was the result, and this was to be followed by some of the world's most famous fighter aircraft. The Mikoyan OKB was established in 1942, with Mikoyan as its manager and Guryevich as chief constructor, and the organisation still exists today as RSK MiG (it was renamed in 1999). Mikoyan died in 1970 and Guryevich in 1976, and in 1971 the reins of the OKB passed into the hands of Rostislav Belyakov who, as general director, was still at the helm in the 1990s.

Pavel Sukhoi learnt his trade as an aircraft designer working for the Tupolev design team. He formed his own OKB in September 1939 and, after the end of the war, became very involved with jet fighter and jet bomber design. In 1945, along with Mikoyan and Guryevich, Lavochkin and Yakovlev, he was commissioned to build jet fighters. Unfortunately, Sukhoi lost favour with Stalin and lost his OKB in 1949 but, after the latter's death in 1953, it was reopened. Sukhoi was responsible for some of the most successful of all Soviet fighter designs and today the AVPK Sukhoi organisation has the job of designing and building the next-generation Russian fighter. After Sukhoi died in 1975, Mikhail Simonov was to become a very successful Chief Designer of the OKB.

Aleksandr Yakovlev was another designer to produce a successful series of wartime piston fighters, his OKB having been one of several to be formed in the late 1930s to respond to Air Force requirements for new aircraft. The lightweight and fast Yak-3 was probably the best Soviet-built pure fighter of the war period and Yakovlev's production success continued with his early jet fighter designs. The growing capability of the MiG and Sukhoi Bureaux however, meant that, apart from the Yak-25/Yak-28 series, he was less successful with his later conventional supersonic designs. In truth that did not matter because the OKB diversified rather more than the others in building short-range airliners and other types while, in terms of fighters, Yakovlev moved on to the first Soviet vertical take-off aeroplanes. The man himself, however, was very politically attuned and aligned with the leadership of the Communist Party, rather more than most of his colleagues, which sometimes allowed him to influence future policy in his favour. In the 1930s he was critical of the famous aircraft designer Andrei Tupolev and played a part in his imprisonment, which resulted in hostility between Yakovlev and parts of the Soviet aircraft industry. Yakovlev died in 1989 but the design team is still active today.

Pavel Sukhoi.

Aleksandr Yakovlev.

Finally, a word about two of the official bodies with whom the OKBs had to work. TsAGI was the short title for the Central State Aerodynamic and Hydrodynamic Institute based at Zhukovskiy. It was formed in 1918 to provide the aviation industry with the results of its scientific research and give technical recommendations for future developments and designs. During the 1920s TsAGI's experimental facilities were expanded and in 1925 it acquired the world's largest wind tunnel. In due course TsAGI's work was supplemented by TsIAM (the Central Institute of Aviation Motors) and VIAM (the All-Union Institute for Aviation Materials) but TsAGI continued to undertake fundamental research into many aspects of aircraft design. A key area from the late 1940s onwards was the very high speeds made possible by the development of ever more powerful jet engines, and the new aerodynamics, materials and structures that were needed to handle them (which forms a core element to the subject matter of this book). Such work required many new facilities and TsAGI had a large input in the great majority of the USSR's fighter designs.

The NKAP or People's Commissariat of Aircraft Industry was formed in January 1939 when it was split away from the People's Commissariat of Defence Industry. The new body was required to improve the organisation, planning and management of aircraft production because there was a growing need to prepare the Soviet Air Force for a possible conflict with Germany and Japan. As such it operated throughout the Great Patriotic War but in 1946, as part of a governmen-

tal shake up, it became the Ministry of Aircraft Industry or MAP. As such it was headed by M V Khrunichev and for the most part led an independent existence within the Soviet military-industrial complex. MAP's leaders had a lot of power in controlling the design and development of new military aircraft and Khrunichev was succeeded by P V Dement'yev in 1953. The latter served for twenty-four years until he was succeeded by V A Kazakov (1977 to 1981), I S Silayev (1981 to 1985) and finally A S Systov, who held the post until the collapse of the USSR in 1991.

Such was the control during the post-war years that the OKBs could be chastised for working on non-authorised projects. The centrally planned nature of Soviet aircraft procurement made for a climate where, for example, access to wind tunnels was dependent on the official sanctioning of a project. Therefore, unofficial designs would rarely proceed very far without a requirement to fill. In America and Britain a requirement was still vital for new designs to progress, but in those countries there was more scope for private venture designs to proceed much further than was usually the case in the Soviet Union (although the Yakovlev Yak-120 did begin as a private venture before being ordered into production as the Yak-25).

The book *Russian Aviation and Air Power in the Twentieth Century* (Frank Cass 1998) describes the 1950s and 1960s as the 'Golden Age' of Soviet aviation because 'national strategic priorities provided continuing requirements for new military and civilian aircraft and almost unlimited funds to support research, development and production'. However, from the 1970s onwards the Soviet economy was less able to sustain the expense of designing as many new aircraft types as had been possible previously. Competing designs were therefore more often assessed while they were still on the drawing board rather than at the prototype stage, less prototypes were built, and those that were were far more sophisticated and complicated. Towards the end of the 1970s the centralised control that had been held over the aircraft industry began to fall away and progress with new designs became dependent on the decisions of bureaucrats working within the central ministries, with the Design Bureaux lobbying for support. This reduced MAP's ability to integrate the work of the OKBs and the production factories. In due course the fall of the Soviet Union brought the end of the centrally controlled and managed aircraft industry.

However, back in 1945 and having just emerged victorious from the Second World War (which was known to the USSR as the Great Patriotic War), an end to the Soviet Union must have seemed near impossible. The records suggest that during the years following the conflict a high proportion of the fighter designs drawn by Soviet OKBs specialising in this field were actually built, although not all of them were flown. Relatively few designs for unbuilt fighter projects have emerged and Yefim Gordon's book *Early Soviet Jet Fighters* (Midland 2002) covers those that were built in some depth. It was during the late 1940s that the maximum number of Soviet OKBs to work primarily on fighters at any one time, five, were all active. The designs that appeared from Alekseyev, Lavochkin, Mikoyan, Sukhoi and Yakovlev during this period are summarised below, and are related by the fact that almost all of them feature straight wings – the move to more advanced aerodynamic wing shapes will be described in the next chapter. It should be noted that the designations used here for both Sukhoi and Yakovlev were, in many cases but for different reasons, re-used by designs produced some years later.

Alekseyev I-21 Series

Alekseyev I-211 and I-215

During the Great Patriotic War Semyon M Alekseyev had worked closely with Semyon A Lavochkin's Design Bureau, in particular on the La-5 and La-7 piston-powered fighters. In 1946 Alekseyev became Chief Designer at Aircraft Factory No 21 at Gor'kiy and was instrumental in producing, among several different types of aeroplane, a series of jet fighter designs embraced by the overall designation I-21.

On 29th April 1946 an official directive was issued by the Council of People's Commissars for new jet fighters to be designed that were to be powered by jet engines offering more thrust and performance than had previously been available from German units. This was a blanket directive aimed at several OKBs and Alekseyev's response was 'Istrebitel-21', the designation coming from the Factory number. Work on this aeroplane had begun in the spring and an Advanced Development Project was ready during the following August. The aircraft, powered by two Lyul'ka TR-2 engines, was required to show a top speed of 980km/h (609mph) at sea level and 970km/h (603mph) at 5,000m (16,404ft). However, this engine had fallen behind its development schedule and when the first I-21 prototype (the I-211 or 'I-21 version 1') made its maiden flight in the autumn of 1947 (possibly on 13th October), two TR-1s had been substituted. An I-210 project (I-21 version Zero!) had been proposed with two BMW 003 engines, but these offered a top speed that fell short of the requirements. Although it was intended to fit three 37mm cannon in the lower nose, the I-211 prototype appears to have carried only 23mm guns.

Early flight testing showed that the I-211 gave a performance which exceeded the original estimates for a TR-1 installation, and it outperformed the competing Sukhoi Su-11 described below, but problems with the engines plus a landing accident ensured it remained the only example of the type to be built. The manufacturer's flight tests gave the I-211 a service ceiling of 13,600m (44,619ft) and time to 5,000m (16,404ft) of 3 minutes. However, several months before the I-211 flew, work had started on a re-engined version with two British Derwent V units called the I-215, and the opportunity was taken to convert the damaged I-211 airframe. As the I-215, the aircraft made its second first flight on 18th April 1948 and was favourably received, but it fell short of the Air Force's requirements (which themselves had advanced with time) and this meant that there was never any chance of the type entering production or service.

The competing Yakovlev Yak-23 (below), with the same engine, was slightly slower but

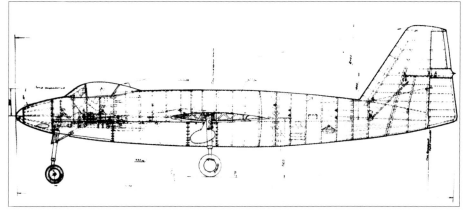

Alekseyev I-21 (1946).

Internal detail for the I-21.

showed a better rate of climb and more agility. However, a second I-215 was built (as the I-215D) to try out a bicycle undercarriage instead of the first machine's tricycle arrangement. The bicycle wheels with outriggers had been proposed for several new aircraft projects, not least the Baade Type 150 jet bomber, and had endured some opposition and criticism, but when the I-215D first flew in October 1949 it proved that the idea would work. In August 1948 Alekseyev's OKB was closed and two months later he became the new head of the Baade OKB-1 working on the Type 150 bomber. Work on his fighter designs, however, came to an end and those prototypes that had flown were subsequently scrapped.

Alekseyev I-212

An altogether different proposal was this two-seat long-range escort fighter design from spring 1947, which was intended to be powered by two VK-1 engines. The project was approved on 14th May and introduced sweepback on its tail and fin but retained the straight wings. The pilot and gunner/radar/radio operator sat in tandem and a Toriy (Thorium) radar set was fitted. In addition, there was a large quantity of internal fuel while a 550kg (1,212lb) drop tank could be carried under each wing – the underwing pylons could also carry an alternative 500kg

Artist's impression of the Alekseyev I-21.

Above: **Alekseyev I-211 prototype.**   Below: **Alekseyev I-215 prototype.**

(1,102 lb) bomb. Laminar flow wings of trapezoid shape were used and, as previously, the two VK-1 engines were housed in wing nacelles. When the I-212 was proposed, however, this power unit was still nowhere near ready for flight test so the OKB's design team switched to the less powerful RD-45 Nene copy as an insurance.

The I-212 was to be an all-metal aircraft which used high-strength steel or V-95 high strength aluminium alloy for those parts of the structure that were required to take heavy loads, although the airframe would also have employed a lot of very light magnesium alloy sheet and components. It was to have two powerful alternative armaments – four Nudel'man Sooranov NS-23 23mm cannon in the nose and two more G-20 20mm cannon in the tail end of the fuselage, or two 23mm plus a single NS-45 45mm cannon in the nose and two 23mm at the rear. It was hoped that the aircraft would meet the Air Force's requirements very closely and, at a normal all-up weight of 9,250kg (20,392 lb), its service ceiling was estimated to be 14,000m (45,931ft) and maximum range (with drop tanks) 3,100km (1,927 miles). The designation UTI-212 was given to a planned trainer version and it was expected that the I-212 would also fulfil night fighter and reconnaissance duties. As noted, during this period and for some time afterwards many fighter prototypes were built and flown, most of which failed to enter production. As a result there was no need to fly another and consequently the I-212 prototype was never finished, although construction of the tailplane at least is believed to have got under way.

Alekseyev I-213 and I-214
Two further proposed versions of the I-212 received new project numbers, despite differing very little from the original and having the same span, length and wing area plus two Nene (RD-45) engines; neither was built. The I-213 was a design study for a heavier version of the '212' and was nearly identical in appearance, although it had just two 23mm guns in the upper nose and one rearwards-facing 23mm in the back end of the fuselage. Normal all-up weight was 10,950kg (24,140 lb) and top speed 1,000km/h (622mph)

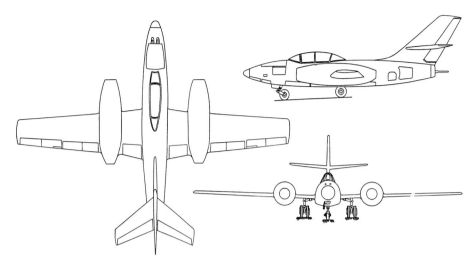

at ground level. The I-214 was also very similar but dropped the tail guns for a radar installation and, to make up for the lack of tail weapons, had two 37mm cannon in the upper nose and a single 57mm cannon in the lower left side of the nose. Estimated top speed was still 1,000km/h at low level but the all-up weight was 10,500kg (23,148 lb); for both versions the time to 5,000m (16,404ft) was 2.7 minutes.

Alekseyev I-216

A development of the I-215, the I-216 project showed no changes to the wings or the fuselage and used Derwent engines, but it was intended to carry two heavy 76mm cannon in the nose. All-up weight would be 7,500kg (16,534 lb), top speed 930km/h (578mph) at ground level and estimated time to 5,000m (16,404ft) 3.1 minutes. Again this design was not built.

Antonov Fighter Project

Antonov Salamandra

The Antonov Design Bureau was opened in response to an NKAP order made on 6th March 1946 and was to become well known for a long series of transport and civilian aeroplanes. However, in its earliest days the new OKB also had a go at fighter design, despite at one stage having its staff numbers reduced by a third. During the war Oleg Konstantinovich Antonov had been Deputy Designer to the Yakovlev OKB, working on the Bureau's series of piston fighters. He was impressed by the German Heinkel He 162 Salamander light-weight 'last-ditch' jet fighter, which was designed and produced in an incredibly short time during the closing stages of the European war. This aircraft had its engine mounted on the top of the fuselage and also twin tail fins to keep the rudders out of the jet efflux, so it is no surprise that the resulting Antonov project shared a pretty similar layout – and it was also called Salamandra. Design work began in the spring of 1947 just after the OKB's team had completed the SKh (An-2) utility transport.

Salamandra had a single RD-10 on the back of its fuselage and this position removed the

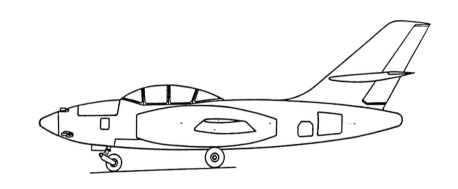

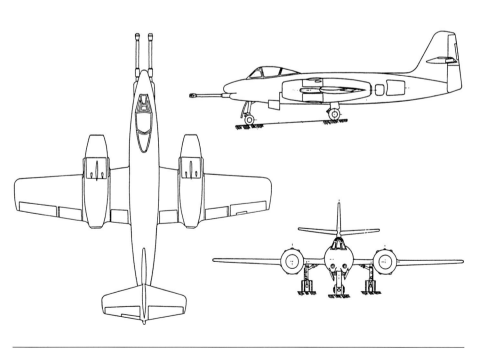

Alekseyev I-216 with two 76mm nose guns (1947).
Russian Aviation Research Trust

Artist's impression of the Alekseyev I-216.

Lavochkin Fighter Projects

Lavochkin 'Aircraft 150'

The OKB led by Semyon A Lavochkin began work on new jet fighter designs in February 1945. Thanks to a new specification being issued to several OKBs at once, Lavochkin was now working in competition with other organisations, which had not been the case when the La-VRD was designed a year earlier. In April the OKB's effort was concentrated on two separate projects, the lightweight 'Aircraft 150' with a single German Jumo 004 engine (actually the Soviet RD-10 copy) and the heavier twin-engined 'Aircraft 160' (below). Once again, in choosing the 150's layout the design team relied on material and data supplied by TsAGI, the aircraft eventually showing a nose intake; a straight one-piece wing placed high on the fuselage and, built integrally with it, a short main fuselage 'pod' with the engine exhausting beneath what in effect was a single boom; and a conventional tail and fin. It also had a tricycle undercarriage and mounted two 23mm cannon, one on each side of the lower forward fuselage. In fact, compared to some early fighter designs, and also several of Lavochkin's later products, the '150' was quite an attractive little fighter.

The advanced development project for the '150' was completed in April 1945, a full-size mock-up was built during May and June, and a pre-production batch of five aircraft was planned. However, because Lavochkin's base at Factory No 81 had limited manufacturing capacity, these aircraft were to be built at Factory No 381 and this brought delays. In fact it took almost a year to build the five aeroplanes because, up to this time, Factory No 381 had previously only worked on wooden aircraft. In addition, following an

risk of ingestion and foreign object damage which came with the rough-field operations for which the fighter was intended; it also made maintenance relatively easy. A model was tested in the wind tunnel but on 6th April 1946 instructions were received from NKAP that the OKB should begin designing another fighter powered by two RD-10s. The Salamandra was dumped and Antonov, who had realised that the performance offered by jet powerplants made the potential success of using unconventional airframe configurations a much stronger possibility, switched to the 'M' or Masha described in Chapter Two.

Above: **Model of the Alekseyev I-216.**

Below: **Model of the Antonov Salamandra (spring 1947).**

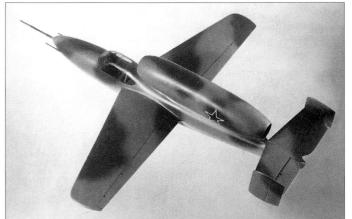

Lavochkin 'Aircraft 150'.

The first Lavochkin '150' prototype as built, but prior to receiving modifications before its first flight.

Lavochkin '150F' prototype with afterburning engine.

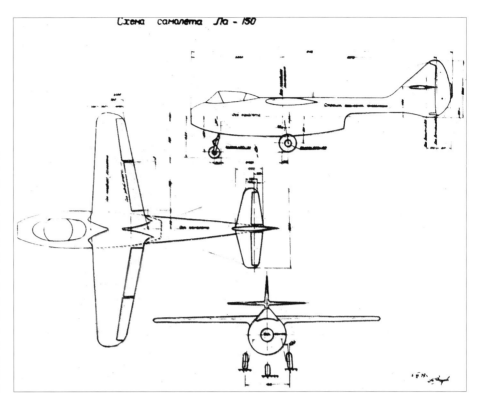

NKAP ruling, in October 1945 the Lavochkin OKB (and plant No 81 as well) moved to Aircraft Factory No 301 at Khimki to allow the OKB to expand its facilities and, from now on, build its own prototypes (NKAP was the People's Commissariat for Heavy Industry). All of this contributed to the '150' programme falling behind schedule and by the end of 1945 only the static test airframe had been finished at Factory 381. The rear fuselage on the static airframe then failed, primarily because the steel portions of its structure had not received the correct heat treatment to increase the material's strength and hardness properties to the required level. Eliminating this problem and repeating the static tests brought further delay and so the opportunity was also taken to modify the wings and tail and increase the fin area.

By July 1946 the first '150' prototype's modifications were complete, but then the first high-speed ground runs showed that during rotation the tail touched the runway which resulted in some ballast having to go into the forward fuselage to push the CofG further forward. Finally, on 11th September the '150' made its first flight. In due course eight aircraft were built, receiving the unofficial service designation La-13, but the type showed numerous weaknesses and was not selected for full-scale production. One example received modifications, including separately manufactured wings, and was redesignated '150M' (M for *modifitseerovannyy* or modified). In this form it flew on 24th July 1947 but a key weakness of the '150' was the unreliable engine, and the powerplant eventually brought a premature halt to the '150M's flight test programme as well. With the higher-performance 'Aircraft 156' now on the way (below), work on the '150M' was abandoned on a decision made by Lavochkin himself.

However, the lack of thrust supplied by the RD-10 did prompt the fitting of a modified unit equipped with afterburning, the aircraft being redesignated '150F'. The second prototype was converted and flew as such on 25th July 1947. It showed considerable improvements in performance, including at top speed at sea level of 950km/h (590mph), and 915km/h (569mph) at 4,320m (14,173ft), which now made the machine one of the Soviet Union's

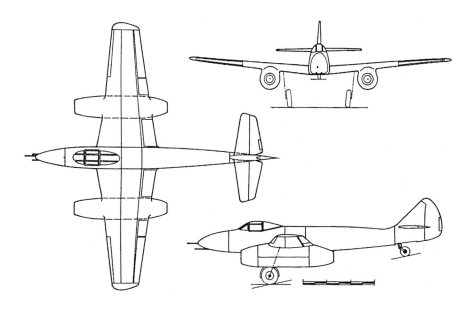

Lavochkin Aircraft '160' (4.45). Russian Aviation Research Trust

cept and destroy heavy enemy bombers. The '160' would have carried 1,223kg (2,696 lb) of fuel, which would have given a reasonably good range and endurance; for example cruising at 850km/h (528mph) and 4,000m (13,123ft) the '160' was expected to be able to stay in the air for half an hour. A tailwheel undercarriage was used and the '160' had three NS.23 23mm cannon mounted in its nose. It soon became clear to Lavochkin that his design team lacked sufficient manpower to work on two new fighters at once. In 1946, after deputy Semyon Alekseyev had departed to Gor'kiy to run his own OKB, Lavochkin shelved the '160' to allow his team to concentrate on the smaller '150' project. Alekseyev had initiated and was the driving force behind the '160'.

Lavochkin 'Aircraft 152'

The next design was intended to be more advanced than the '150'. In fact work began in late 1945 after the '150' mock-up had revealed the size limitations of that design and, although it still used a single RD-10, a larger fuselage was introduced. The wings were moved to a lower position and the engine was now in the forward fuselage, not aft of the cockpit; also, compared to the '150', an extra 23mm cannon was installed on the starboard side of the nose. The prototype first flew on 5th December 1946 but the bigger aircraft, coupled with the same power unit, did show a small loss of performance. Just the one example was built and this was not repaired following damage suffered in a landing accident in mid-1947.

Lavochkin 'Aircraft 154'

In late summer/early autumn 1946 Lavochkin began an effort to improve the '152's performance by installing a TR-1 engine instead of the RD-10. There were no other structural changes and the OKB began building a prototype in September, but the engine's delivery was held up until the stage was reached where there was no longer any point in completing the aircraft. Therefore the project was abandoned. Some sources state that the '154' was expected to have a sea level rate of climb of 1,590m/min (5,215ft/min), take 3.7 minutes to reach 5,000m (16,404ft) compared to the measured 5.6 minutes of the '152', and have

fastest jet fighters. However, the weaknesses remained, one of which was that the slim fuselage had been designed purely to take the RD-10 powerplant, so there was no room to fit a larger engine. Lavochkin moved on to 'Aircraft 152' described shortly.

Lavochkin 'Aircraft 160'

This project represented the first use of the '160' designation (the second was a swept-wing fighter described in Chapter Two). Powered by twin RD-10s, it was intended to be larger and heavier than the '150' and to serve as an insurance for the latter because it was thought that the '150's powerplant might not give sufficient thrust for that aircraft to offer an adequate performance. However, two RD-10s did not really give enough thrust for the '160' either and so this design's armament and fuel were cut back to compensate. This move reduced the type's combat potential, although its objective was essentially to inter-

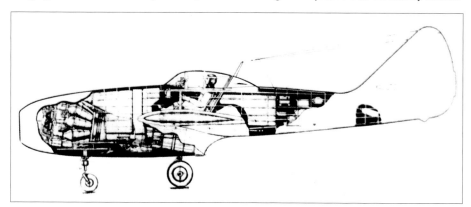

Cutaway drawing of the Lavochkin '152'.

Heavily retouched photograph of the '152' showing the nose gun arrangement, and the aircraft's rather ugly shape.

The Lavochkin '154' prototype never received its powerplant and so was never completed.

a service ceiling of 12,850m (42,158ft). However, there is some conflict because a recent Russian-language book on the history of Lavochkin states that the '154' was to be powered by a single 8.8kN (1,985 lb) RD-10 engine giving a take-off weight with just two 23mm cannon in place and 730kg (1,609 lb) of fuel aboard of 3,000kg (6,614 lb). This less powerful engine would also produce an estimated top speed of 950km/h (590mph) at sea level and 1,000km/h (622mph) at 5,000m (16,404ft), take three minutes to reach 5,000m and give a service ceiling of 14,000m (45,932ft) and a range of 600km (373 miles).

Lavochkin 'Aircraft 156'

This was really begun as a purpose-built testbed aircraft for the RD-10 engine fitted with reheat and a Special Design Bureau within the Lavochkin OKB was responsible for designing and developing the afterburner. The whole power unit was designated 'Arti-

cle YuF' (for uprated Jumo 004) and the '156' itself was actually a version of the '152' fitted with reheat, so to begin with the aircraft was designated '152D' (D for *dooblyor* or second prototype). There were some other changes to the structure, including new wings and greater tail and fin area, and work began on 22nd November 1946. The first of two prototypes flew on 1st March 1947 and reheat was

used for the first time for take-off on 10th April. The '156' took 4.2 minutes to reach 5,000m (16,404ft), and mock combat with a MiG-9 showed that the extra power gave the '156' a big advantage. However, the prototypes still suffered from plenty of problems and, following a critical official report issued in January 1948, further work on the development of the '156' was closed.

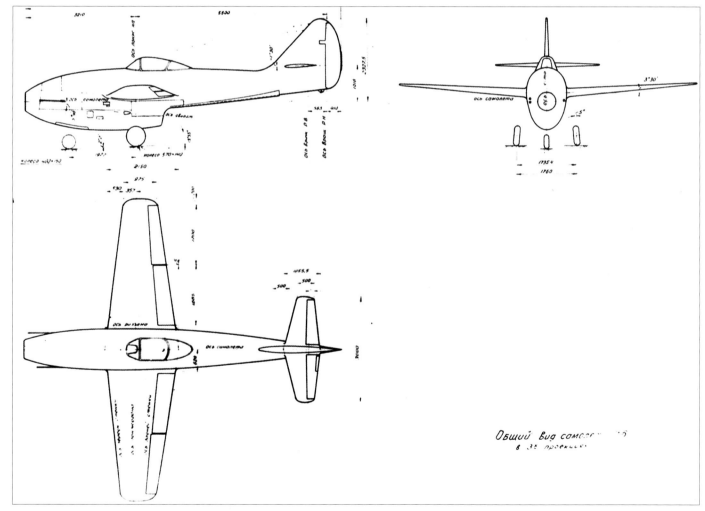

Lavochkin '156' (11.46).

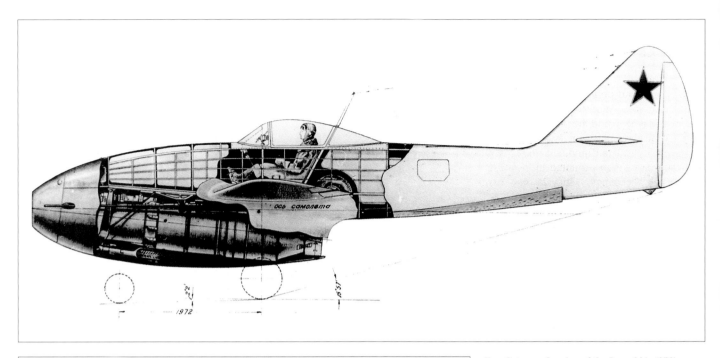

Lavochkin 'Aircraft 174TK'

There was one further development of Lavochkin's straight-wing fighter family. The '174TK', rather confusingly numbered out of sequence, was a further development of the '156' but with a thinner wing (TK = *tonkoye krylo* or thin wing). In fact, when flown with a t/c ratio of 6%, this aircraft had what is believed to be the thinnest wing yet flown on a fighter. However, when the '174TK' became airborne in January 1948, from a state of the art point of view it was already well behind the swept-wing 'Aircraft 160' (see Chapter Two) and so it remained a prototype. Its performance data included taking 2.5 minutes to reach 5,000m (16,404ft), a service ceiling of 13,500m (44,291ft) and a range of 960km (597 miles).

Mikoyan MiG-9 'Family'

Mikoyan I-260

During February 1945 the Council of People's Commissars issued a directive that requested several OKBs to design and construct new single-seat jet fighters. We have seen the results that this generated at Lavochkin, and another OKB to become involved was Mikoyan. The resulting I-260 project was, like Alekseyev's I-21 and Sukhoi's first jet fighters, greatly influenced by the German Me 262, which had two engines in underwing

Below: **The prototype Mikoyan I-300 jet fighter.**

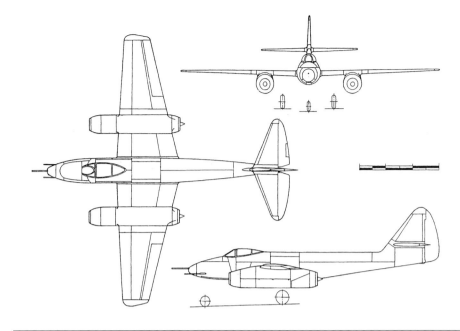

nacelles and had performed so well during World War Two. Work began on the I-260 design in May 1945 and the project was completed at the end of the year. It was also called 'Izdeliye F' or 'Article F' and had a heavy battery of three cannon mounted in the nose, two BMW-003 engines, a tricycle undercarriage and straight wings, but there was substantial leading-edge sweep on the horizontal tail and vertical fin. Span was approximately 9.9m (32ft 6in) and length including the nose guns 10.1m (33ft 1½in). However, from a drag point of view and the need to make maximum use of all of the available engine power, Mikoyan's design team felt that it could do better. In due course the two engines were moved to the lower forward fuselage and the resulting project, the I-300, initiated in June 1945 just a month after the I-260 showed that this alternative arrangement offered some advantages. The I-260 was subsequently dropped.

Mikoyan I-300 and MiG-9

The I-300 project, or 'Izdeliye F' (the designation being re-used), shared a roughly similar layout to the Lavochkin series described above (the so-called 'pod-and-boom' arrangement) but had two BMW 003 engines placed side-by-side in the lower fuselage and

was altogether bigger and heavier. The I-300's main duty was to deal with heavy bombers, for which it was to carry either one 57mm or one 37mm cannon, plus two 23mm guns, all mounted in the nose (the 57mm was fitted in the intake splitter). The I-300 project was approved in late autumn 1945 and a full-size mock-up was built. It was the first Soviet fighter to use a tricycle undercarriage. Work on the prototype was held up because the aft fuselage and tail had to be strengthened following recommendations received from

TsAGI concerning the strength of high-speed airframes. The first machine finally flew on 24th April 1946, just a few hours before Yakovlev's Yak-15 (below), making it the Soviet Union's first indigenous jet-powered aircraft to fly.

The I-300 also became the first Soviet jet aircraft to enter full production, entering service as the MiG-9 (the West allocated the codename *Fargo*). Various improvements and upgrades followed, including the MiG-9M (I-308) with reheated RD-21 engines first

flown in July 1947, the new engines giving a maximum thrust of 9.8kN (2,204 lb) and a top speed of 942km/h (585mph) at 3,000m (9,842ft). However, after its state trials the MiG-9M prototype was rejected for several reasons, but by then the more advanced swept-wing I-310 (the MiG-15 prototype, Chapter Two) was under way and so any further work on this straight-wing project would have been a waste of time.

Mikoyan I-320

The arrival of British jet engines in Russia gave Mikoyan the opportunity to fit a single Nene power unit into the basic MiG-9 airframe. The resulting project was called the I-320 (and 'Izdeliye FN') and the construction of a prototype began in late 1947. However, again because more advanced swept-wing types offering better performance were just around the corner (and the I-310 used the same engine), the project was abandoned and the airframe was scrapped before completion. On the MiG-9 itself, trouble had been experienced with the guns inducing gas ingestion into the engines and several alternative positions were tried out. On the I-320 it had been intended to put the 37mm gun into the lower left nose (without having the muzzle ahead of the intake) and one 23mm either side of the nose (again set back from the intake).

Sukhoi Research Project

Sukhoi Jet Engine Powered Aircraft

The Sukhoi OKB's first essays into jet aircraft design were actually made well before the end of World War Two, in fact in this case as early as October 1942, although the design

Artist's impression of the Mikoyan I-320.

never left the drawing board. It makes an interesting comparison with its contemporaries described above. At this time Pavel Sukhoi's efforts were concentrated on increasing the top speed of Soviet fighters, which included some attempts to produce piston fighters fitted with a booster jet engine (such as the Su-5 flown in April 1945).

This rather odd project, which was really a research aircraft design with no armament, was an all-metal design with an annular air 'intake' scoop surrounding the fuselage behind a teardrop-shaped cockpit. The nose section housed both the cockpit and a fuel tank and was attached to the larger diameter central fuselage section on four pylons; the cutaway drawing shows this arrangement well – although it looks somewhat precarious. The central fuselage contained the pow-

erplant, which was actually made up of an air-cooled engine with an oil cooler and twin propellers that served as an air compressor, followed by the jet engine's fuel injection section and finally the combustion chamber made in the form of a tapered tube. The wing had two spars and comprised a centre section, built integrally with the fuselage, and two outer panels. There were two wing fuel tanks. The elevators had trim tabs and the whole tailwheel-type undercarriage could be retracted.

Sukhoi Jet Engine Powered Aircraft (10.42).

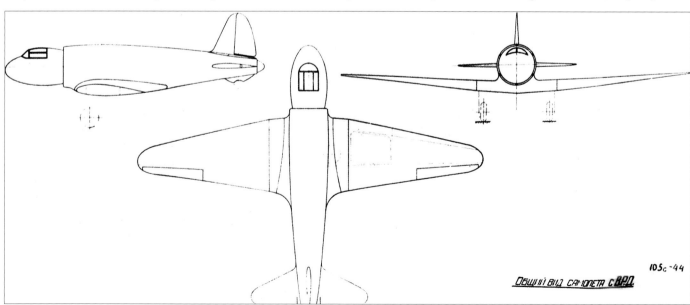

Soviet Secret Projects: Fighters Since 1945

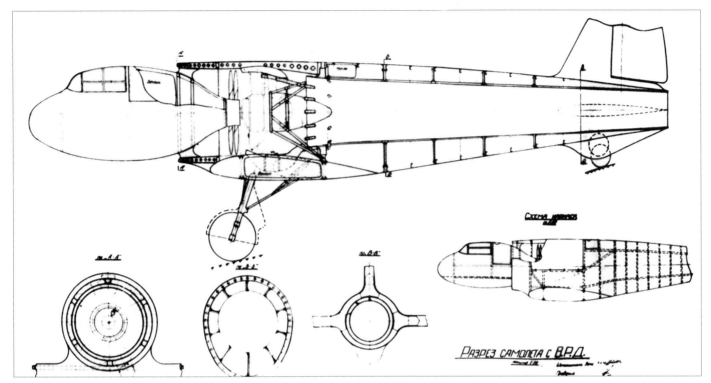

Sukhoi Su-9 'Family'

Sukhoi Su-9

Also known as 'Izdeliye L' (Su-9 was not allocated until later), this project was started during 1944 and was accepted within the Soviet aircraft industry's development line-up for 1945, although on several occasions during that year it was removed from the programme and then reinstated. This was another aircraft to resemble the Me 262 – it was not a copy but was clearly influenced by the German machine (Mikoyan's I-260 was a lot closer to the 262's shape). Unlike Mikoyan, Sukhoi preferred the underwing engine position and stuck with it, and the 'L' featured an ejection seat, rocket bottles to reduce the take-off run and was the first Soviet combat aeroplane to use powered controls. Initially it was intended to fit two Lyul'ka S-18 (TR-1) engines, but delays resulted in the German RD-10 (004B) being substituted instead.

A mock-up was officially examined on 7th February 1946 but the 'L' fell behind its schedule, not least because the Sukhoi OKB also had to undertake design work on the trainer variant of the piston-powered Tupolev Tu-2 light bomber. Changes in the recommendations received from TsAGI regarding the use of laminar flow aerofoils also meant that the fighter's wings were redesigned three times. Nevertheless, when rolled out the fighter, now redesig-

nated Su-9 and also 'Article K', was actually only about three months late and made its maiden flight on 13th November 1946. In general it performed well and after its state trials the type was recommended for production, but by now orders had been placed for other jet fighter types and so the K/Su-9 remained the only prototype, ending its flying career in May 1948 for economic reasons. It was scrapped, although Sukhoi had wanted to present the airframe to the test pilot school based at the LII Research Institute. The Su-9 achieved a service ceiling of 12,800m (41,995ft) and took 4.2 minutes to get to 5,000m (16,404ft).

A second prototype had been started and it was intended to fit reheat to the engines

Sukhoi Jet Powered Aircraft cutaway.

which, as with other OKBs working on adding this feature to their aircraft, required the design work to be done in-house. In the case of the Su-9 of course, the work needed to be done to two separate nacelles. However, this was achieved together with alterations to the wingtips and re-aligning the engine axes with the wing chord, which also required further changes to the structure. Estimates suggested that, with afterburning RD-10Fs, the aircraft would now achieve 925km/h (575mph) at sea level and 908km/h (564mph) at 6,000m (19,685ft). The airframe reached an advanced

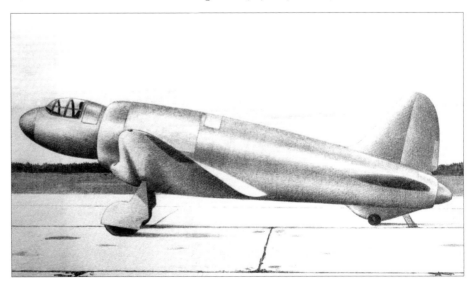

Artist's impression of the Sukhoi Jet Powered Aircraft.

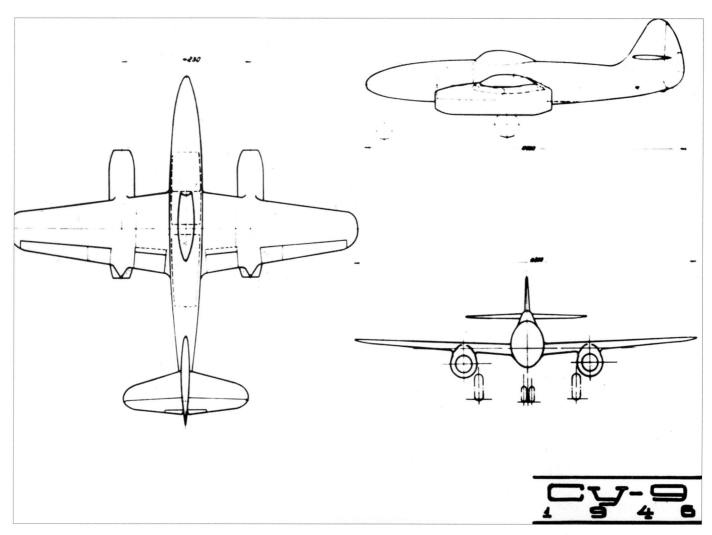

CV-9
1 9 4 6

Sukhoi Su-9 (1946).

This model, believed to be the Su-9, was displayed on the Sukhoi stand during a 1990s Farnborough Air Show.

stage of construction before a Council of Ministers directive dated 29th April 1946 halted work and requested that TR-1 powerplants should be fitted instead. In this form the aircraft became the Su-11.

Sukhoi Su-11

Since the Su-9 was first proposed as a TR-1-powered fighter, things had come full circle. The airframe and guns were unchanged from the Su-9, only the Su-11's engine nacelles and position on the wing had been altered to any visible degree. The Su-11 (also known as 'Izdeliye LK' for aircraft 'K' fitted with Lyul'ka engines) made its first flight on 28th May 1947, but the engines gave insufficient power, preventing the fighter from reaching its specified performance. In addition it suffered from a lack of longitudinal stability at high speeds plus several other problems, and so was soon

abandoned. By the end of April 1948 the aircraft had been scrapped but its wing went to TsAGI for static structure testing. In flight the Su-11 showed only a slight increase in ceiling over the Su-9 to 13,000m (42,651ft) but took just 3.6 minutes to reach 5,000m (16,404ft).

Sukhoi Su-13

Another member of the family, the 'KD'/Su-13, was created in the summer of 1947 when the TR-1s were replaced again, this time by the British Derwent V. This gave the chance to alter the wings further, to reduce the t/c ratio from 11% to 9%, to carry drop tanks beneath the wingtips and to give the

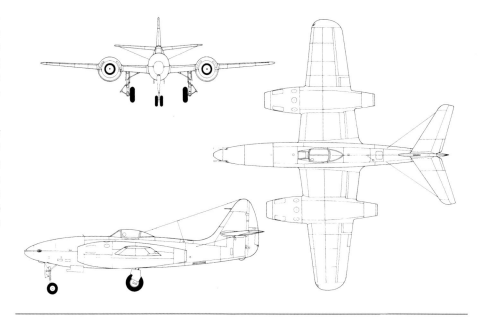

Top: **This was the only prototype of the Su-9 to be built.**

Above: **Sukhoi Su-11 prototype.**

Right: **Sukhoi Su-13 (1947).**

horizontal tail 35° of sweepback. About 20% of the wing tooling and wing detail was completed and two Su-13 prototypes were to be built, but MAP refused to authorise them and the project had been terminated by summer 1948. An all-weather interceptor version of the Su-13, still a single-seater and known as the 'TK', was also proposed with a Toriy (Thorium) radar scanner in the nose in between two 37mm guns, plus two RD-500 Soviet copies of the Derwent V. Length was increased to 11.9m (39ft 0½in) and all up weight to 6,757kg (14,896 lb). Top speed was estimated to be 980km/h (609mph) at 4,500m (14,764ft) and both Su-13 variants were expected to take 2.5 minutes to reach 5,000m (16,404ft). The preliminary project for the all-weather Su-13 was completed in March 1948 but was not presented to MAP for appraisal.

Yakovlev Fighter Projects

Yakovlev Yak-15

Prior to 1945 the Yakovlev Design Bureau's experience with jet propulsion was rather limited, but on 9th April 1945 it was ordered to build a single-seat jet fighter powered by the Jumo 004 engine. Time was saved by using the piston Yak-3 fighter's airframe as a starting point, which had the power unit placed underneath the forward fuselage with the nozzle just about level with the pilot's seat – in other words the 'pod and boom' arrangement again. Designated provisionally Yak-Jumo or Yak-3-Jumo, the only major structural change from the Yak-3's wing was that the front spar was made U-shaped to pass above the engine.

The prototype was ready in October 1945 but Yakovlev's inexperience showed through when, during engine runs, the lower fuselage aluminium skinning beyond the engine heat shield melted and the rubber on the tailwheel (which was right in line with the exhaust) burned. Changes had to be made (including a solid steel tailwheel which showered the runway with sparks) and these took until late December to complete. Finally, following extensive wind tunnel testing, the first prototype made its maiden flight on 24th April 1946, just a few hours after the MiG I-300. Flight-testing went well and in due course the two prototypes were refitted with Soviet-built copies of the 004 (the RD-10). As such the fighter was ordered into production as the Yak-15, the degree of commonality with the piston Yak-3 ensuring that the type's real purpose was to fill the gap between piston fighters and the newer jet fighters, designed from scratch, which would follow. A total of 280 were built and the Air Standards Co-ordinating Committee called the type *Feather*.

Yakovlev Yak-17 (First)

A more improved jet fighter was requested from Yakovlev by a Council of Ministers directive dated 29th April 1946. The resulting machine was based on the Yak-15 but showed many changes, not least redesigned wing and tail surfaces for higher speeds. In fact two different wings were considered which had respective areas of 13.5m² (145.2ft²) and 15.0m² (161.3ft²), the smaller giving the higher estimated top speed of 850km/h (528mph) at 7,000m (22,966ft) compared to the larger machine's 830km/h (516mph). With the larger wing the estimated service ceiling was 14,000m (45,931ft). The wing itself was too thin to accommodate the mainwheels and so the main legs had to be redesigned and these were attached to one of the fuselage mainframes. Prototype construction moved ahead very well and a first flight was pencilled in for 20th September 1946 but, in the event, the Yak-17 prototype never flew. Ground testing was halted on 26th September and, because the Yak-15 had been ordered in quantity, no further examples were built. The designation was almost immediately re-used. Theoretically the thinner wings would have given this aircraft a modest increase in speed over the Yak-15 powered by the same engine but the latter

The sole Yak-17 prototype with a tailwheel undercarriage was never flown, but the designation was re-used.

The Yak-17 designation was re-allocated to a modification of the Yak-15 with a tricycle undercarriage.

had been ordered into production; the tailwheel landing gear was also outdated.

Yakovlev Yak-17 (Second)

Alongside the development of the Yak-17 above, the OKB began work on a revised Yak-15 with a tricycle undercarriage called the Yak-15U-RD-10. This was flown in June 1947 and was eventually selected for production in March 1948 as the Yak-17. This type shared the *Feather* codename.

Yakovlev Yak-23

On 11th March 1947, after British-built Nene and Derwent jet engines had become available, the USSR Council of Ministers issued a new directive requesting several OKBs to design new aircraft powered by these engines. Yakovlev was ordered to do a tactical straight-wing fighter, which became the Yak-25 below, but the designer decided to do two projects at once, his second 'private venture' project retaining the 'pod and boom' format. It was designated Yak-23 and the prototype flew in July 1947, and it looked an altogether larger and beefier aircraft than its predecessors. Flight trials were successful and in late 1948 the type was ordered into production as an insurance against problems and possible failure of the swept-wing Mikoyan MiG-15 and Lavochkin La-15 (Chap-

Prototype of the Yakovlev Yak-23.

ter Two). The ASCC's recognition codename for this aircraft was *Flora*.

Yakovlev Yak-19

The Yak-15 and Yak-17 fighters showed that the RD-10 gave insufficient power to provide a fighter that was fully combat-capable, so directives were issued for new fighters to use the Lyul'ka TR-1. However, progress with this powerplant was slow and therefore, as a stop-gap, Yakovlev opted to uprate the RD-10 with reheat. After the new version of this engine passed its tests on 4th May 1946, the OKB began a new fighter design called the Yak-19, which had to be completed in a very short time. It was, construction taking just four months, and the prototype flew on 8th January 1947 as the first Soviet aircraft to be fitted with afterburning (although the Lavochkin '156' operated its reheat before the Yak-19's was first lit on 21st May 1947).

The Yak-19 was quite different from its pre-decessors. The old pod-and-boom layout with the powerplant in the forward fuselage gave way to a more aerodynamically efficient arrangement with the engine buried in the aft fuselage and exhausting via a long extension pipe, the first time this was used in Soviet design practice. It also had a tricycle undercarriage and a second prototype flown on 6th June had tip tanks but the availability of better-quality British engines prevented the construction of a third machine, so the project was halted by the OKB itself. During its state acceptance trials the Yak-19 showed a service ceiling of 12,100m (39,698ft) and, in afterburner, took 4.0 minutes to achieve 5,000m (16,404ft). In flight the aircraft flew well and it became the first Soviet aircraft to achieve 900km/h (559mph).

Yakovlev Yak-25

At this stage it seems rather odd that another straight-wing fighter should have been planned

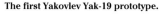

when swept-wing types like the MiG-15 were now in the air. Probably more than any other fighter OKB leader, Aleksandr Sergeyevich Yakovlev had a close relationship with the Soviet Government and its establishments, his opinions carried weight and he still regarded swept wings with some caution. As a result, Yakovlev's 'official' response to the 11th March 1947 directive was a design which had a straight wing and an all-through jet pipe like the Yak-19, but also a swept tail and fin. It retained the Yak-19's tricycle undercarriage and had three 23mm cannon in the lower nose behind the intake.

The Advanced Development Project for this aircraft, the Yak-25, was approved by Yakovlev on 1st February 1947 and the first prototype was ready in the following October. After such terrific progress the fighter made its maiden flight on the 31st of that month. Three exam-

ples were built and, during its state acceptance trials, the type displayed a much higher top speed (up to 50km/h or 31mph more, dependent on height) and ceiling over the Yak-23, despite a higher all-up weight. The fighter met its specification in full but its performance still lagged behind the MiG-15, so there was never any likelihood of production. Yakovlev had to accept that straight wings were out of date and the OKB moved on to the swept-wing designs described in Chapter Two.

Yakovlev Yak-27

This single-seat fighter project was designed in 1947 and had a layout that was similar to the Yak-25; in other words it was a straight-wing design having the wings in the mid position but with swept tail surfaces. The powerplant comprised a single Nene in the rear of the fuselage and in its initial form, as

The first Yakovlev Yak-19 prototype.

first drawn in February 1947, this aircraft featured a circular nose intake. However, in May it was redesigned with wing root intakes along the lines of the Yak-30 trainer aircraft (not the Yak-30 described in Chapter Two). There are few details but wing area was 18.7m² (201.1ft²) and estimated take-off weight 4,150kg (9,149 lb). This project did not proceed beyond its initial studies.

–

To close this chapter, it must be noted that throughout 1945, a period representing the genesis of Soviet jet fighter design, the Soviet government became extremely frustrated at the slow pace of development shown by pretty well all of the OKBs. As a result new directives were issued, either by the government or the State Defence Committee, on an almost monthly basis that gave ever more precise tasks and deadlines, but they made little difference (only a few have been mentioned in the above text). In truth, it was very difficult to adapt factories and facilities to absorb this new technology within a certain time. The task of creating a 'jet fighter industry' would have to, and did, take as long as required, despite the ever-higher blood pressures suffered by those in government. As will be seen in later chapters, following their 'conversion' to jet technology, the 'Fighter OKBs' would prove to be very successful.

The first Yakovlev Yak-25 prototype.

Straight-Winged Jet Fighters – Data / Estimated Data

Project	Span m (ft in)	Length m (ft in)	Gross Wing Area m² (ft²)	All-Up-Weight kg (lb)	Engine kN (lb)	Max Speed / Height km/h (mph) / m (ft)	Armament
Gudkov Gu-VRD	10.5 (34 5.5)	9.0 (29 6.5)	11.0 (118.3)	2,250 (4,960)	1 turbojet 14.7 (3,305)	870 (541) at S/L, 900 (559) at 6,000 (19,685)	1 x 20mm cannon, 1 x 12.7mm (0.5in) machine gun
Lavochkin La-VRD	9.80 (32 2)	9.00 (29 6)	15.0 (161.3)	3,300 (7,275)	1 x S-18 c12.2 (2,755)	890 (553) at S/L, 850 (528) at 5,000 (16,404)	2 x 23mm cannon
Alekseyev I-21 Project	12.25 (40 2)	11.54 (38 0)	25.0 (268.8)	7,600 (16,754)	2 x TR-2 22.0 (4,960)	900 (559) at 6,000 (19,685)	3 x 37mm, or 1 x 37mm + 2 x 23mm cannon, 2 x 250kg (551 lb) bombs
Alekseyev I-211 (flown)	12.25 (40 2)	11.54 (38 0)	25.0 (268.8)	8,010 (17,658)	2 x TR-1 12.7 (2,865)	950 (590) at S/L, 910 (566) at 4,000 (13,123)	2 x 23mm cannon
Alekseyev I-215 (flown)	12.25 (40 2)	11.54 (38 0)	25.0 (268.8)	7,850 (17,306) estimated	2 x Derwent V 15.6 (3,500)	970 (603) at S/L, 947 (589) at 4,000 (13,123)	3 x 37mm cannon
Alekseyev I-212	16.20 (53 2)	13.08 (42 11)	32.8 (352.7)	10,500 (23,148)	2 x RD-45 (Nene I) 22.2 (5,000)	1,000 (622) at S/L, 960 (597) at 8,000 (26,247)	See text
Lavochkin '150' (flown)	8.20 (26 11)	9.42 (30 11)	12.15 (130.6)	2,973 (6,554)	1 x RD-10 8.8 (1,985)	840 (522) at S/L, 878 (546) at 4,200 (13,779) '150F' with reheat: 950 (590) at S/L, 915 (569) at 4,320 (14,173)	2 x 23mm cannon
Lavochkin '160'	11.00 (36 1)	10.20 (33 6)	20.2 (217.2)	4,020 (8,862)	2 x RD-10 8.8 (1,985)	850 (528) at 5,000 (16,404)	3 x 23mm cannon
Lavochkin '152' (flown)	8.20 (26 11)	9.04 (29 8)	12.15 (130.6)	3,160 (6,966)	1 x RD-10 8.8 (1,985)	840 (522) at 5,000 (16,404)	3 x 23mm cannon
Lavochkin '154'	8.20 (26 11)	9.42 (30 11)	?	3,490 (7,694)	1 x TR-1 12.2 (2,755)	925 (575) at S/L, 920 (572) at 5,000 (16,404)	2 x 23mm cannon
Lavochkin '156' (flown – Manufacturer's flight test data)	8.52 (27 11.5)	9.12 (29 11)	13.32 (143.2)	3,521 (7,762)	1 x RD-10YuF 11.5 (2,580)	926 (576) at S/L in reheat, 882 (548) at 5,000 (16,404) rh	2 x 23mm cannon
Lavochkin '174TK' (flown)	8.64 (28 4)	9.41 (30 10.5)	13.52 (145.4)	3,315 (7,308)	1 x Derwent V 15.6 (3,500)	970 (603) at S/L, 965 (600) at 3,000 (9,842)	3 x 23mm cannon
Mikoyan I-300 Prototype (flown)	10.0 (32 10)	9.75 (32 0) without guns	18.2 (195.7)	4,988 (10,996)	2 x RD-20 7.8 (1,765)	920 (572) at 5,000 (16,404)	1 x 57mm, 2 x 23mm cannon (57mm replaced by 37mm on MiG-9)
Mikoyan I-320	10.0 (32 10)	10.88 (35 8)	18.2 (195.7)	?	1 x Nene 22.2 (5,000)	?	1 x 37mm, 2 x 23mm cannon
Sukhoi Su-9 (flown)	11.2 (36 9)	10.545 (34 7)	20.2 (217.2)	5,890 (12,985)	2 x RD-10 8.8 (1,985)	847 (526) at S/L, 900 (559) at 3,000 (9,843)	1 x 37mm, 2 x 23mm cannon, 2 x 250kg (551 lb) bombs
Sukhoi Su-11 (flown)	11.8 (38 8.5)	10.549 (34 7)	21.4 (230.1)	6,277 (13,838)	2 x TR-1 12.7 (2,866)	890 (553) at S/L, 904 (562) at 4,000 (13,123)	1 x 37mm, 2 x 23mm cannon, 2 x 250kg (551 lb) bombs
Sukhoi Su-13	11.8 (38 8.5)	10.93 (35 10)	24.8 (266.7)	6,436 (14,189), max 7,036 (15,511)	2 x RD-500 15.6 (3,500)	960 (597) at S/L, 960 (597) at 5,000 (16,404)	3 x 37mm cannon, 2 x 250kg (551 lb) bombs
Yakovlev Yak-15 (flown)	9.2 (30 2)	8.7 (28 6.5)	14.85 (159.7)	2,570 (5,665)	1 x RD-10 8.8 (1,985)	770 (479) at S/L, 800 (497) at 5,000 (16,404)	2 x 23mm cannon (not fitted on first prototype)
Yakovlev Yak-17 (first)	9.1 (29 10)	8.4 (27 7)	15.0 (161.3)	2,700 (5,952)	1 x RD-10 8.8 (1,985)	780 (485) at S/L, 830 (516) at 7,000 (22,966)	2 x 23mm cannon
Yakovlev Yak-17 (second) (Yak-15U-RD-10 prot – flown)	9.2 (30 2)	8.78 (28 10)	14.85 (159.7)	3,240 (7,142)	1 x RD-10 8.8 (1,985)	702 (436) at S/L, 751 (467) at 4,250 (13,944)	2 x 23mm cannon
Yakovlev Yak-23 (flown)	8.73 (28 8)	8.1 (26 7)	13.7 (147.3)	3,389 (7,471)	2 x RD-500 15.6 (3,500)	925 (575) at S/L, 910 (566) at 5,000 (16,404)	2 x 23mm cannon
Yakovlev Yak-19 (2nd prot) (flown)	8.72 (28 7)	8.357 (27 5)	13.56 (145.8)	3,400 (7,495) with tanks	1 x RD-10F 10.8 (2,425) reheat	907 (564) at 5,250 (17,224) clean	2 x 23mm cannon
Yakovlev Yak-25 (flown)	8.88 (29 2)	8.66 (28 5)	14.0 (150.5)	3,580 (7,892) with tanks	2 x RD-500 15.6 (3,500)	950 (590) at S/L, 972 (604) at 3,000 (9,842)	3 x 23mm cannon

Chapter Two

Swept Wings and Transonic Performance

Alongside the rest of the world, the Soviet jet fighter began to make real progress when it adopted swept wings. As noted in Chapter One, the first swept-wing designs were under way well before some straight-winged types had been completed, so the transition took several years. Once again, a high percentage of these new designs were turned into hardware and flown.

Alekseyev Projects

Alekseyev I-211S and I-217
A version of the Alekseyev I-211 described in the last chapter was also proposed with wings, tail and fin all swept; it also featured a longer fuselage and the project was called I-211S. Approximate dimensions were span 13.0m (42ft 8in) and length 13.2m (43ft 3¾in).

The last fighter proposals from the Alekseyev OKB, before the Bureau was closed in August 1948 and Alekseyev moved on to the Baade OKB, also featured swept wings. In fact there were two versions of the I-217, the second very advanced in that it had its wings swept forward, a feature rarely used throughout any period of aviation history. This last pair of projects, which are believed to have got going during 1947, were again essentially swept-wing developments of the earlier I-211/I-215 series, I-217 meaning 'I-21 Version 7'. The swept-back variant was an attractive aircraft that had a span of 12.25m (40ft 2in) and length 11.59m (30ft 0½in) – no other data is available but it did retain the 'straight' tail and fin. Both I-217s had three cannon in the lower nose.

The forward-sweep version was a bit bigger, and this fascinating aircraft was to be powered by two Lyul'ka TR-2 engines. It was

Lavochkin 'Aircraft 160'.

estimated that the time required to reach 5,000m (16,404ft) would be 1.5 minutes, and to 10,000m (32,808ft) 4.0 minutes, and that the estimated range was 1,900km (1,181 miles). During an interview made during the early 1990s Semyon Alekseyev declared that his Design Bureau never had any interest in building this forward-swept-wing fighter project – essentially it was just an experimental study.

Antonov Lightweight Fighters

Antonov 'M'
In 1947, following the efforts made on the Salamandra (Chapter One), the Antonov OKB moved on to a study for a lightweight single-

seat tailless 'flying wing' fighter. On official instructions it was to be powered by two RD-10 engines and these were to be fed by root intakes. The project's wing had 60° of sweep on the leading edge and had full span flaps plus all-flying, forward-swept ailerons, the latter forming the outer wing sections. Twin fins were employed, mounted at the tips inside the ailerons, and the armament comprised a quartet of 23mm cannon (or alternatively two 37mm and two 23mm), one each housed underneath the intakes and two in the nose. The aircraft had a tricycle undercarriage and was named Masha, or abbreviated as 'M'; its chief designers were A A Batumov and V A Dominikovskiy.

The design was nearly complete when, late in 1947, the OKB was told to replace the RD-10s with an RD-45 (Nene) and redesign the aeroplane as required. In fact, apart from the forward fuselage, the redesign was total, but after tunnel testing the construction of a prototype proceeded with some haste. The revised layout had a new wing of larger span and area (in fact a delta wing with rounded

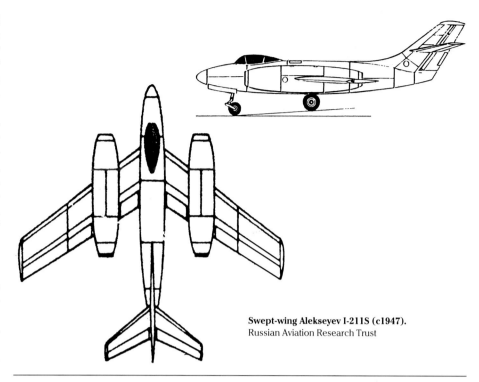

Swept-wing Alekseyev I-211S (c1947).
Russian Aviation Research Trust

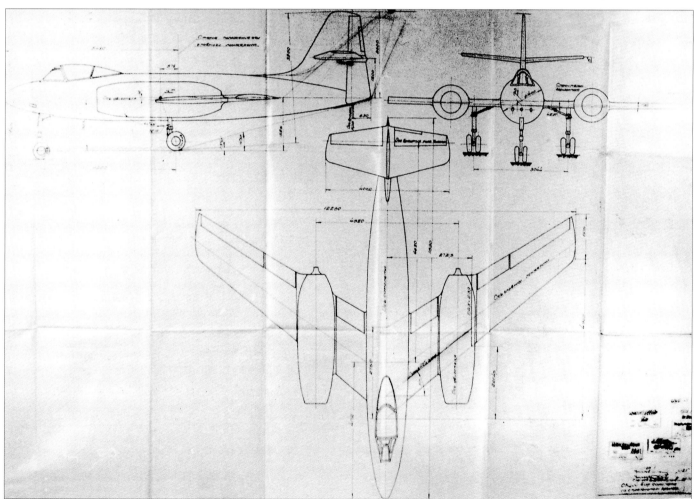

Alekseyev I-217 with back-swept wing (1947/48).

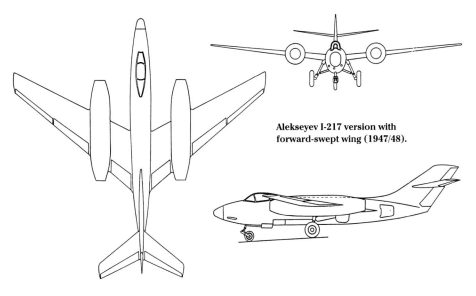

Alekseyev I-217 version with forward-swept wing (1947/48).

tips), different root intakes and the tip ailerons were replaced by elevons. The estimated data included a time to 5,000m (16,404ft) of five minutes, ceiling 10,000m (32,808ft) and a range when flying at 8,000m (26,247ft) of 620km (385 miles). A 1/10th-scale flying model was produced to check the aircraft's flight and spinning characteristics and some of the performance estimates, but it proved impossible to check every mode of flight and so the decision was made to construct a full-scale model glider.

Antonov E-153

This was the designation given to the full-size glider, which was built in wood and also served as a structural and equipment mock-up. The aircraft was to be launched using a dolly as an undercarriage, to be jettisoned after take-off, while a fixed skid was fitted for landing. In July 1948, shortly before the glider's flight test programme was due to start, the 'M' and E-153 programmes were terminated on MAP orders, official opinion being that the new fighters now coming from Lavochkin, Mikoyan and Yakovlev would be sufficient. At this stage the construction of the prototype Masha was nearly finished and Antonov was disappointed with the decision because he thought that the 'M' would have been far more manoeuvrable than any of its contemporaries – he felt that it really should have flown. There had also been the potential, in due course, to add a radar and a more powerful engine.

Baade OKB Projects

Junkers EF 126

By the end of the War a number of German aircraft and engine factories and their design facilities had been overrun by the Soviet army and many of their engineers and designers captured. A good number of these people were taken back to Russia and put to work on new aircraft and powerplants, where they were allowed to make use of their knowledge in advanced aerodynamics, jet engines and structures. In due course two new Design Bureaux were established at Podberez'ye and these were staffed with both German and Russian engineers working alongside one another. OKB-1 was officially formed on 22nd October 1946 and was led by Brunolf Baade and Hans Wocke. It worked mostly on bombers but also continued with the pulse-jet-powered Junkers EF 126 fighter and pro-

Model of the forward-swept I-217. Neither version of this aircraft was built.

Soviet Secret Projects: Fighters Since 1945

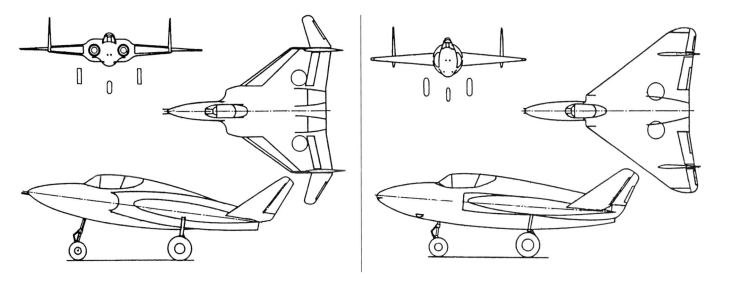

Above left: **Original layout of the Antonov 'M'**
(1947).

Above right: **Definitive layout of the Antonov 'M'**
(1948).

Right: **Sketch showing the original Antonov 'M'.**
Russian Aviation Research Trust

Below right: **Drawing of the Antonov E-153 glider.**
Russian Aviation Research Trust

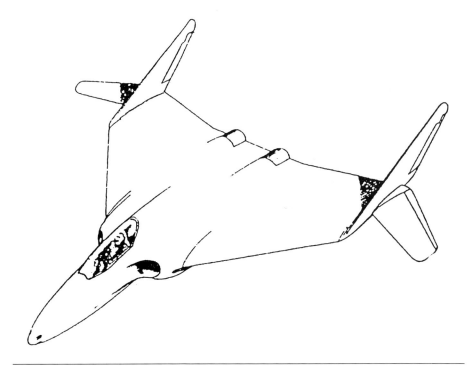

duced the all-new EF 137 project. OKB-2 was
also formed in October 1946 and was to be
employed on bombers as well, but in addition
it put a lot of effort into designing rocket-powered fighters and this work is described in
Chapter Seven.

The EF 126 began life as a fighter project
and formed part of Germany's desperate
attempts to stave off defeat. It resembled and
had much in common with the wartime V-1
unmanned flying bomb except that the war-
head in the nose was replaced by a cockpit.
Soviet officials quickly decided to continue
the project as an experimental aircraft and in
October 1945 work began on five prototypes.
The first of these flew on 12th May 1946 from
Dessau as a glider, having been towed into
the air without an engine. In September three
prototypes were moved to Stakanovo (now
Zhukovskiy) and in March 1947 the gliding
flights resumed but, because of a variety of
development problems, their pulsejet
engines were never brought to a suitable
standard for a full flight test programme to
proceed. The first prototype had been fitted
with the standard flying-bomb engine, the
Argus 109-014 rated at 3.43kN (772 lb) thrust
at sea level, but subsequent aircraft had the
109-044/Junkers Jumo 226 rated at 4.90kN
(1,102 lb). Despite prolonged testing this

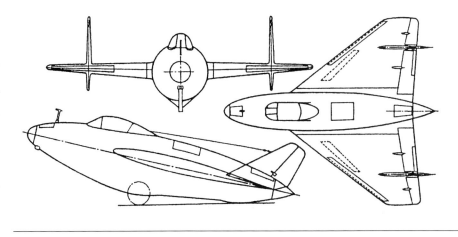

Antonov E-153 glider.

engine suffered from bad ignition, poor combustion and dangerous fires and so the EF 126 was abandoned in 1948.

Baade EF 137
This single-seat fighter was drawn in September 1946 and showed, for its time, a very advanced design with 49° of sweep on the leading edge and variable sweep on the trailing edge. There were ailerons on the reduced-sweep outer wing and the wing itself appears to have been quite thick. The powerplant, two Jumo 004 turbojets, was mounted in an unusual double 'nacelle' beneath the rear fuselage and was fed by a large single ventral intake. A massive cannon,

probably a 37mm, was fixed in the nose, two more guns, probably 23mm, were either side of the cockpit and a tricycle undercarriage was used with the main wheels retracting into the fuselage ahead of the intake. The estimated performance figures included taking 4.75 minutes to reach 6,000m (19,685ft) and 13.75 minutes to get to 10,000m (32,808ft), plus a rate of climb at sea level of 27m/sec (88.5ft/min) and at 6,000m of 15.7m/sec (51.5ft/min). No information is available regarding the fate of the EF 137 but it seems likely that there was no attempt to build a prototype. In 1953 the Baade design team was relocated to Dresden.

Lavochkin Series

Lavochkin 'Aircraft 160'
This was the second use of this designation and covered a type that proved to be the Soviet Union's first aeroplane to use swept wings. It also retained the pod and boom arrangement used by Lavochkin's straight-wing types described in Chapter One, plus

Very poor quality view of the Junkers EF 126 'Elli' pulsejet-powered fighter. Helmut Walther via Russian Aviation Research Trust

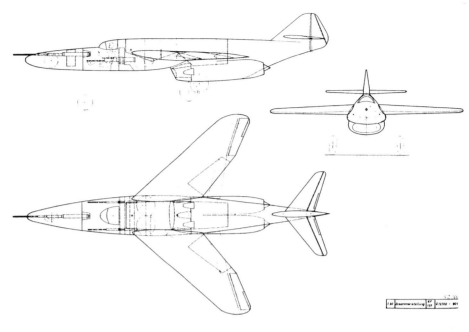

Baade EF 137 (3.10.46). Brunolf Baade and Olaf Baade via Dr Reinhard Mueller

the German Jumo 004 engine. The '160' made its maiden flight on 24th June 1947 but the 8.8kN (1,985 lb) thrust engine did not deliver enough power to get the '160' airborne at maximum weight. Consequently the engine was fitted with reheat which then allowed the '160' to show almost immediately the higher speeds that were possible through having a swept wing. However, the rest of the airframe was essentially an old design and the type was treated mainly as a research aircraft, so there was never going to be any production. The '160' took 4.4 minutes to reach 5,000m (16,404ft), it had a service ceiling of 12,200m (40,026ft) and a range of 1,000km (622 miles); it was nicknamed *Strelka* or *Dart*.

Lavochkin 'Aircraft 168'

The pod and boom were finally discarded by the next prototype, which had a conventional fuselage and tail jet pipe plus a larger swept wing. It also used a single British Nene power unit and had a single 37mm cannon and two 23mm in the lower nose. The first flight was made on 22nd April 1948 but, although probably the best jet fighter design to come out of the Lavochkin OKB so far, the competing MiG-15 (below) was ordered into production. Flight-testing lasted until February 1949 and

Lavochkin 'Aircraft 160'.

the '168' showed that a high subsonic speed was possible with a relatively small amount of thrust. Service ceiling was 14,570m (47,802ft) and the aircraft could reach 5,000m (16,404ft) in 2.0 minutes.

Lavochkin 'Aircraft 174D' and La-15

A scaled-down version of the '168' was produced, with a Derwent 5 replacing the Nene, plus a smaller wing and an extra 23mm cannon instead of the 37mm. As such the project

ran alongside, and formed a comparison to, the straight-wing '174TK' in Chapter One. Some older published sources have called the first prototype the '172', with the second machine being redesignated '174D'. Flight-testing began in late August 1948 and resulted in the design being put into small-scale production as the La-15. Compared with the prototypes the La-15 introduced anhedral on the wing, reducing the span slightly to 8.904m (29ft 2½in), and a stretched fuselage 9.56m

(31ft 4½in) in length, while in service the aircraft's top speed dropped slightly to 1,026km/h (638mph) and the all-up-weight rose to 3,850kg (8,488lb). The La-15 used the RD-500 engine (the Derwent copy) and the ASCC called it *Fantail*. Frustratingly for Lavochkin, it was to prove to be the only production jet fighter to come from a Design Bureau whose piston-engined fighters had been so successful during the Great Patriotic War.

Lavochkin 'Aircraft 176'

There is no better indication of the incredible number and range of fighter prototypes built by the Soviet Union during the late 1940s than the series created by the Lavochkin OKB. The next and last in this group was 'Aircraft 176'. This was an 'improved' version of the '168', which had been a successful aeroplane, and introduced a 45°-sweep wing instead of the 35° angle on the '160' or the 37° 20' of the '168'. The basic configuration of the '168' was retained. For his team's aircraft designs Semyon Lavochkin stressed particularly the need for high speed ahead of other flight parameters and qualities. This is believed to have been the first occasion that a wing of such high sweep angle had been fitted to an aeroplane, and the decision to adopt such an angle created a good deal of discussion and dispute, and doubts, about the wisdom of using it.

The '176' was designed as an interceptor but was intended primarily to serve as a transonic research aircraft. The first flight was made in late 1948 and on 26th December the '176' was taken to a speed of Mach 1.02 in a shallow dive, starting from 10,000m (32,808ft) with a gradual transition into horizontal flight, and this was the first occasion that a Soviet-built aircraft had been taken through the sound barrier. An official report noted that the maximum speed obtained was 1,105km/h (687mph) at 7,000m (22,966ft), and added that the 'data collected in the '176' programme was of exceptional value for our

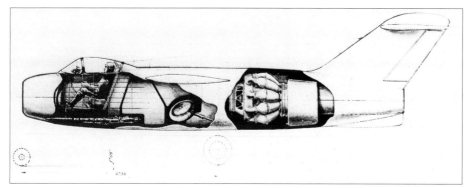

Lavochkin 'Aircraft 168' prototype.

Cutaway drawing of the '168' with its Nene engine clearly shown.

The prototype Lavochkin 'Aircraft 174D'.

Soviet Secret Projects: Fighters Since 1945

aviation (industry)'. At this time this figure was higher than the internationally recognised air speed record. In due course the RD-45 (Nene) engine was replaced by a more powerful VK-1, the introduction of which, together with a slightly higher all-up-weight, removed the problems of flutter that had affected the '168' at high subsonic speeds. However, the sole '176' prototype was lost during a test flight in 1949 when the aircraft's cockpit canopy failed. The '176's service ceiling of the '176' was 15,000m (49,213ft) and the time to 5,000m (16,404ft) was 1.8 minutes.

Despite such progress, with the '176' leading the way in breaking the 'Soviet sound barrier', the MiG-15 showed a better overall performance and the follow-on MiG-17's position was well entrenched from a production point of view; consequently the Lavochkin fighter could do nothing else to push them out of the picture. Nevertheless, it was to be 1949 before a MiG fighter (a MiG-15 fitted with a 3,380kg VK-1F engine) could match the '176's achievement, while the Yakovlev Design Bureau had to wait until 1950 for its Yak-50 (Chapter Four) to achieve a supersonic speed.

Above right: **This view of a production La-15 shows the anhedral given to the wing, which did not feature on the '174D'.**

Right and below: **Lavochkin 'Aircraft 176'.**

Mikoyan Swept-Wing Fighters

Mikoyan I-310 and MiG-15

On 11th March 1947 a Council of Ministers directive was made ordering Yakovlev to produce a new design of jet fighter, the results from this being the straight-wing Yak-23 and Yak-25 detailed in the previous chapter. That same directive also gave Mikoyan the job of producing two prototypes of a fighter that, according to the operational requirement, had to have a top speed of 1,000km/h (622mph) at sea level and 1,020km/h (634mph) at 5,000m (16,404ft). The problem of achieving this speed along with the other stated requirements, such as 3.2 minutes to 5,000m, powerful armament and easy maintenance, prompted Mikoyan to introduce swept wings – a big step forward. During this period the Central State Aerodynamic and Hydrodynamic Institute (TsAGI) carried out a lot of research into the aerodynamics of wings swept back to an angle of 35°, and this eventually was the wing recommended for, and adopted by, the La-160 above and Mikoyan's new fighter.

Several preliminary development projects were put together for possible alternative configurations. One to fall by the wayside early on was a twin-engined type that was rejected principally because it was felt that the Soviet aircraft engine industry would be unable to produce enough engines, within a reasonably short time period, to satisfy the demands of mass production. In addition a twin-boom arrangement was also considered, but this was eventually rejected for a conventional fuselage with a tail jetpipe; the latter, with swept wings and tail surfaces, presented a simple solution that also happened to be the most aerodynamically efficient. The new project was designated I-310.

In July 1951 an American technical aeronautical magazine published an article written by Mikhail Guryevich called *I Designed the MiG-15* that, considering the restrictions and security prevalent in the Soviet Union at that time, was a considerable scoop. Assessments by American experts suggested that most of the content was pretty accurate. The text is fascinating, although it was coloured by the usual rhetoric, such as 'all credit must

go to Marshal Stalin for his inspiration and great wisdom in foreseeing the need for these things and for providing the tools with which we have carried out our work'.

The article noted that the Mikoyan OKB had actually begun its studies for a new swept-wing fighter back in 1946 and these initially considered both forward sweep and sweep-back. The original swept-back design had a straight rudder, the tailplane was sat on top of the fin and there was also a long tailpipe that created losses in engine efficiency. As a result, on the final 'design' this had been altered and the aft fuselage had been undercut to shorten the tailpipe. There was also a swept-back rudder and the tailplane had been moved slightly down the fin. There were problems with the CofG from having the engine in the rear fuselage and this was eventually cured in two ways. First the power unit was moved further forward and, second, the trailing edge wing root was filled in so that, for just a short section next to the fuselage, the wing trailing edge was straight.

'Comrade Christianovich' provided the Design Bureau with the relevant data for air-

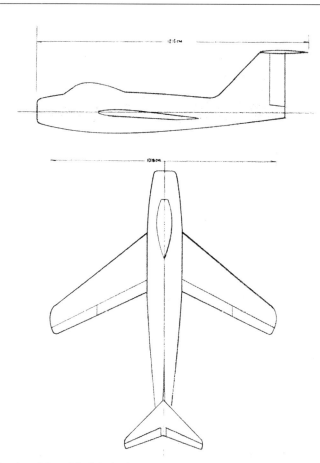

Drawing of the original design for the project leading to the Mikoyan MiG-15, with a straight rudder and long tailpipe (1946). Span c10.1m (33ft 1½in), length 12.15m (39ft 10½in).

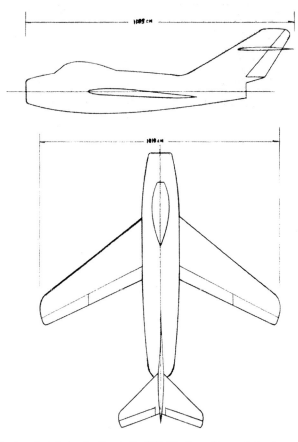

The final project design leading to the MiG-15 with undercut aft fuselage and a swept rudder. Span 10.1m (33ft 1½in), length 11.05m (36ft 3in).

Mikoyan I-310 (MiG-15) first (above) **and second** (right) **prototypes.**

flows at and approaching supersonic speeds which, from the point of view of increases in drag, was vital to the new fighter's development. In addition, tests had shown that using relatively small sweep angles would give little or no performance benefit, the highest possible angles were needed to get the most effective results. However, this then raised the question of creating a wing with sufficient strength so, as a result, the wing loadings had to be kept to a minimum to avoid structural failures. It was after the discussions on this aspect that the designers settled on the 35° sweep angle.

Another problem was the available engine power. At the start Mikoyan had hoped that an engine giving 29.4kN (6,614 lb) would soon be available, but in 1946 it was clear that nothing above 19.6kN (4,409 lb) would be forthcoming within the next twelve months, and so the design team had to adjust their performance estimates accordingly. However, during the prototype's construction the first Nene engines, which now offered at least 21.56kN (4,850 lb) of thrust, arrived from England and the new design was amended to accommodate the foreign engine. As such the OKB was now able to make other changes that brought within reach the top speed that had been the original objective all along.

The I-310 prototype made its maiden flight on 30th December 1946 and its trials were a

success. It was adopted for production and duly entered service as the MiG-15, a designation that was to become world-famous through the type's exploits during the Korean War in the early 1950s. In addition it made the Mikoyan Design Bureau itself world-famous to the point that today, and for many years past, the world's media almost always refer to Soviet or Russian jet fighters as MiGs, regardless of the fact that many of them actually come from Sukhoi or Yakovlev. The ASCC initially allocated the codename *Falcon*, although this was quickly changed to *Fagot*, and the MiG-15 was built in huge numbers and many versions. These included the MiG-15*bis*P or '*Izdeliye* SP-1' which had a VK-1 engine and a large radome fitted to the

upper lip of the intake to take a Toriy (Thorium) radar. Toriy had a single antenna and was specifically intended for the MiG-15 because its compactness meant that no major changes would be needed in the aircraft's structure to take it. The SP-1 prototype flew on 23rd April 1949 but the type did not enter service, primarily because the radar was unreliable (also see Chapter Four). The upgraded MiG-15*bis*, which had introduced the more powerful VK-1, first flew on 22nd July 1949 as the '*Izdeliye* SD'.

Mikoyan I-330 and MiG-17

Apart from the big selection of MiG-15 versions produced during that fighter's career, there was to be a more advanced develop-

ment that was essentially an all-new aircraft. The MiG-15's layout was undoubtedly a good base on which to build. The more powerful VK-1 of the MiG-15*bis* was now combined with a new 45°-sweep wing to satisfy an early 1949 Council of Ministers directive which called for an improved MiG-15 in two forms: a tactical fighter and a radar-equipped all-weather interceptor.

The tactical fighter was designated I-330, or '*Izdeliye* SI', or initially MiG-15*bis* 45. In terms of introducing the new higher-sweep wing the I-330 came behind the Lavochkin '176',

but by now Mikoyan's position as a supplier of jet fighters was well-entrenched and, as noted above, Lavochkin could not beat the latest MiG product. There were quite a number of other changes to the airframe such as increased wing area, a longer fuselage and a new horizontal tail; the aircraft also made extensive use of a new aluminium alloy called V95, but the control system and armament were practically unchanged. Like the MiG-15, for strike operations the I-330 had to be able to carry 100kg (220 lb) bombs on the underwing hard points that were normally

A Mikoyan MiG-17F, an upgraded MiG-17 fitted with a VK-1F engine with afterburning.

used for drop tanks. The first flight was made on 14th January 1950 and the I-330 showed almost immediately that it was faster than the MiG-15*bis*. The changes made from the MiG-15 eventually brought the new service designation MiG-17 and, once again, the fighter was built in large numbers. The ASCC allocated the reporting name *Fresco* and the later MiG-17F used a VK-1F engine fitted with reheat, which gave 33.1kN (7,450 lb) of thrust.

Again other versions followed including the SP-2, another attempt by Mikoyan to fit radar into one of its fighters. This was actually the all-weather interceptor prototype requested by the 1949 directive and it had a Korshoon (Kite) radar. The prototype flew in March 1951 but the radar was too complex for the single crewman to operate and this, coupled with other problems, prevented any production.

Mikoyan SN

There were also prototypes of the SN variant of the MiG-17 that introduced new side intakes (the first Mikoyan fighter to have this feature) and an extended solid nose with movable cannon armament. The novel installation, designed by N I Volkov, had three

Mikoyan SN prototype.

Soviet Secret Projects: Fighters Since 1945

23mm Afanas'yev/Makarov AM-23 cannon mounted asymmetrically, one to starboard and the other pair one above the other on the port side, and was principally intended for ground strafing. The SN prototype was completed on 20th July 1953 and its flight trials continued well into 1954. However, it was found that when the three guns were all angled up or down and fired together the recoil caused pitch up or pitch down. In addition, gun angles greater than 10° prevented accurate shooting while the new intake position did not suit the engine, surging being one of the resulting problems. There was a drop in top speed of some 60km/h (37mph) and manoeuvrability, rate of climb and ceiling were all reduced. The SN was therefore abandoned.

Mikoyan I-340

Also known as the SM, this was yet another prototype built to test an alternative arrangement to the basic MiG-17 airframe. By 1950 the new Mikulin AM-5 axial engine had become available giving 19.6kN (4,409 lb) of thrust from a diameter around half that of the VK-1. Since two of these units together would give more thrust than one reheated VK-1, Mikoyan took the opportunity to sit a pair of them side by side in the rear of one of the SI/MiG-17 prototypes. The AM-5 was intended to be the new engine that would power some planned future fighters from Mikoyan and Yakovlev, as outlined in a Kremlin meeting held on 30th June 1950 and attended by Mikoyan, Yakovlev, Mikulin and Minister of Aviation Industry M V Khrunichev.

The I-340's targets were to provide an improvement in performance over the MiG-17 and to gain experience with the AM-5 so that good reliability and fuel economy could be established. A wider rear fuselage was needed to accommodate the new power units, the whole of the inlet ducting had to be enlarged and a ventral strake was fitted. The project was approved by the Council of Ministers on 20th April 1951, preliminary design began in May and the prototype was ready by the end of the year. It was rolled out in March 1952 and flown soon afterwards, but the power supplied by the AM-5s, even with reheat added in the AM-5A to give 21.1kN (4,740 lb) of thrust, could not push the aircraft right through the sound barrier on the level,

although Mach 1.19 was achieved in a shallow dive. As a result Mikulin modified the AM-5's compressor and produced a new afterburner to create a more powerful engine called the AM-9. The I-340 took 0.94 minutes to get to 5,000m (16,404ft) and had a service ceiling of 15,600m (51,181ft).

Sukhoi Tactical Fighter

Sukhoi Su-17

Following the straight-wing designs discussed in Chapter One, the next project to come from Pavel Sukhoi's Design Bureau was the swept-wing radar-equipped Su-15 described in Chapter Four. This was followed by an aeroplane that was intended to

be supersonic on the level, to undertake research at high subsonic, transonic and low supersonic speeds, and to act as a prototype for a tactical fighter. Known in-house as 'Aircraft R', it fell within the plans issued by the Council of Ministers in late 1947 for experimental aircraft for the 1948-49 period and was ordered alongside the Yakovlev Yak-1000 (Chapter Three). Funding was given for one prototype and one static test airframe of each type, and this was a relief because government cuts had removed several other projects from the programme and closed down a number of aircraft and engine Design Bureaux.

In 1949 the Soviet Air Force (VVS) redesignated Sukhoi's aircraft Su-17. A full-scale mock-up was examined and approved in

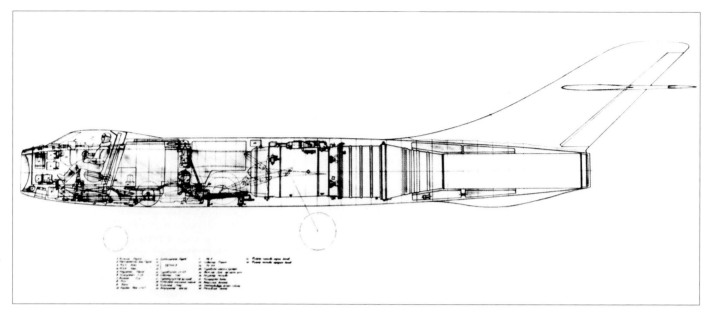

Sukhoi Su-17 (1948).

December 1948 and the prototype was completed during the summer of 1949, without the two heavy cannon intended to go in the fighter version. The 'R'/Su-17 had a 50° swept wing plus a forward fuselage which, in an emergency, could be jettisoned as a whole with the pilot and cockpit inside. The prototype arrived at LII, the Flight Test Research Institute at Zhukovskiy, in July 1949 and by early August had begun its high-speed taxi tests. Then, when the aircraft was ready for its first flight, an order was received forbidding the Su-17 to take-off.

In the Soviet Union a new aircraft was not allowed to fly for the first time without the approval of TsAGI and here the Institute's research scientists had calculated that the Su-17's swept wing was weak and might suffer from flutter at high speeds. It was thought that the wing's single spar might not provide enough rigidity, flight clearance was never

given and the Su-17 never did get airborne. The loss of the Su-15 prototype during a test flight also became a contributory factor, Sukhoi being blamed by TsAGI for the crash. The Su-17 prototype was returned to the Design Bureau for some strengthening to be added into the wing, but then Stalin ordered that Sukhoi's OKB should be dissolved on 1st November. Pavel Sukhoi rejoined the Tupolev OKB's design team, his engineers were reassigned elsewhere and in due course the Su-17 went back to the research centre at Zhukovsky to be used in the ground testing of crew rescue systems. After that the airframe was sent to a firing range to serve as a target for aircraft cannon.

Thus it was never confirmed that the Su-17 would be supersonic, but the performance estimates indicated a maximum speed at sea level of 1,209km/h (751mph) or Mach 0.985 for the fighter and 1,252km/h (778mph) or

Mach 1.022 for the research prototype; respective figures at 10,000m (32,808ft) altitude were 1,156km/h (718mph) Mach 1.1 and 1,152km/h (716mph) Mach 1.07, service ceilings 14,500m (47,572ft) and 15,500m (50,853ft) and times to 10,000m (32,808ft) 4.4 and 3.5 minutes. A key element of this potentially outstanding aeroplane was the new TR-3 engine from Lyul'ka, which was later redesignated AL-5 but at this stage was not fully developed. More powerful engines could have been fitted in due course. When Sukhoi's facility was closed the OKB had about thirty different projects on-going for jet-powered aircraft, including transports and civil airliners and the 'N' or Su-14 two-seat attack aircraft, the prototype of which at this stage was about 40% complete. Another casualty was an escort fighter powered by a Klimov VK-2 engine that was expected to show a range of 5,000km (3,108 miles).

Yakovlev Series

Yakovlev Yak-29

This was a small single-seat lightweight fighter project studied between July and October 1947 and powered by a single Derwent engine mounted in the rear of the fuselage. The intake was located above the fuselage in the dorsal position behind the cockpit and the Yak-29's mid-position wing was initially given a sweep angle of 35°, but this was later increased to 45°. Length was

The Su-17 was a sleek and attractive fighter but was denied the chance to show its profile in the skies.

8.64m (28ft 4in), span 6.3m (20ft 8in), wing area 8.0m² (86.0ft²) and all-up weight 2,300kg (5,071 lb), but it is not known as to which wing arrangement these figures apply. This data gave a wing loading of 287kg/m² (58.97 lb/ft²), which was a rather high figure. Versions with two cannon or no armament were studied.

Yakovlev Yak-30

The first Yakovlev type to feature swept wings, this aircraft was derived from the Yak-25 (Chapter One). It introduced 35° sweep at the leading edge and was powered by a single RD-500 engine, the Soviet copy of the British Derwent. The first of two prototypes was flown on 4th September 1948 and the type displayed a good combination of high maximum speed (Mach 0.935) and excellent manoeuvrability in both horizontal and vertical planes, although there were problems including poor aileron compensation and using an incorrect wing aerofoil section. But in truth the Yak-30 was too far behind the MiG-15 to make any real impression and, after its manufacturer's flight-testing was concluded in mid-December, the fighter was not submitted for official testing. Service ceiling was 15,000m (49,213ft) and time to 5,000m (16,404ft) 2.6 minutes.

A second machine, known as the Yak-30D, had a 38cm (15in) plug inserted into the rear fuselage, modified mainwheel doors, Fowler flaps and other changes, which increased the normal all-up weight by 110kg (242.5kg). It joined the flight test programme in early 1949 and the modifications showed that the designers had solved some of the first machine's problems; however, such progress was by now irrelevant.

Yakovlev Yak-40 and Yak-41

These designations were applied to variants of the same basic single-seat design with a 45° sweep wing, work on them beginning in January 1948. The main feature was that they were to be powered by a pair of 8.3kN (1,875 lb) ramjet engines mounted at the wingtips, in nacelles 0.55m (1ft 9⅝in) in diameter and 2.5m (8ft 2½in) in length. The Yak-40 had a length of 7.50m (24ft 7in) and an all-up weight of 1,800kg (3,968 lb). Two cannon were mounted in the fuselage nose and take-off assistance was available in the form of solid fuel rockets, which on the Yak-40 were fixed to a trolley that was to be jettisoned after take-off. The fighter was intended to operate

in combination with a carrier aircraft converted from a Tupolev Tu-4 bomber. Up to six Yak-40 fighters were to be suspended under the bomber's outer wings, the fighter's wings overlapping to allow stowage of three of these small aeroplanes within a restricted space beyond each of the Tu-4's outer engines. The Yak-40 designation was later reused for a passenger aircraft.

The alternative Yak-40A had two 7.35kN (1,655 lb) thrust U-93-1 rocket motors fixed beneath the wing, replacing the trolley. The Yak-41 was a concurrent experimental variant with a 2.94kN (661 lb) liquid-fuelled rocket in the fuselage tail; its span was 4.5m (14ft 9in), length 7.9m (25ft 11in) and take-off weight 2,300kg (5,071 lb). Research work on these aeroplanes was halted in June 1948 and none of them progressed beyond the drawing board. This type of layout with tip-mounted engines has been examined for use on fighter types by several aircraft design teams around the world. In general, it was rejected – a big

weakness being the problem of asymmetry when one engine had been lost.

Yakovlev Yak-M

Derived from the earlier radar-equipped Yak-50 (Chapter Four), this lightweight single-seat fighter introduced side-fuselage intakes placed level with the pilot's cockpit. It was to be powered by a single AM-5 jet and in the normal load condition had just the two fuselage fuel tanks, holding 350kg (772 lb) and 250kg (551 lb) respectively; however, for overload a further 200kg (441 lb) tank could be added between the first two. The performance estimates made use of the drag data collected from the Yak-50, but modified to take into account the M's different layout. These indicated a sea-level rate of climb of 3,600m/min (11,811ft/min), service ceiling 17,500m (57,415ft) and a flight endurance in the normal load condition when flying at 5,000m (16,404ft) of one hour forty-five minutes; the latter could be increased to three

hours in the overload state. Work on the Yak-M project lasted from November 1950 until February 1951 and the aircraft may have been a competitor to the Mikoyan I-340.

Yakovlev Yak-U

Two fighter projects shared this designation, the first a variant of the Yak-M which was under way during April and May 1951 as the Yak-U (L). The second, another 1951 project but unrelated to the first, appears to have been a larger aircraft. It was still a single-seat fighter but, like the concurrent Mikoyan I-340, was to be powered by a pair of AM-5 engines placed side-by-side in the fuselage. The fuselage itself was of oval cross section, had a

nose intake (in an arrangement similar to the Mikoyan MiG-19) and a bicycle undercarriage, and also housed three fuel tanks, holding 925kg (2,039 lb), 150kg (331 lb) and 225kg (496 lb) respectively. Neither of these projects were built.

Yakovlev Yak-70

Yet another single-seat fighter study, this time fitted with a Lyul'ka TR-3. This engine, also used by the Sukhoi Su-17, was designed between 1946 and the beginning of 1947 and used a seven-stage axial-flow compressor to gave 4,600kg (10,141 lb) of thrust. An improved version designated TR-3A was produced in 1948 that gave a maximum thrust of

4,615kg (10,174 lb), and in 1952 the power was increased again to 5,100kg (11,243 lb). At the beginning of the 1950s the engine was manufactured in small numbers as the AL-5, but in the event it was to be installed only in prototype aircraft. Work on the Yak-70 was short-lived, lasting through just April and May 1950, and the fighter had a nose intake with a central shock cone for a radar. A bicycle (zero-track) undercarriage was fitted, which was to become the Yakovlev OKB's trademark, and the fuselage had a diameter of 1.60m (5ft 3in) and was 15.24m (50ft 0in) long. No performance figures are available but it can be assumed that transonic speeds would have been achieved.

Swept-Winged Jet Fighters – Data / Estimated Data

Project	Span m (ft in)	Length m (ft in)	Gross Wing Area m² (ft²)	All-Up-Weight kg (lb)	Engine kN (lb)	Max Speed / Height km/h (mph) / m (ft)	Armament
Alekseyev I-217 (forward sweep wing)	14.20 (46 7) 45.1 (10,150)	13.85 (45 5.5)	31.0 (333.3)	8,550 (18,849)	2 x TR-2	1,100 (684) at S/L	3 x 37mm cannon
Antonov M (mid-1947)	10.8 (35 5)	10.6 (34 9.5)	?	?	2 x RD-10 8.8 (1,985)	?	4 x 23mm cannon
Antonov M (late 1947)	9.3 (30 6)	10.64 (34 11)	?	?	1 x RD-45 22.2 (5,000)	950 (590) at height	4 x 23mm cannon
Baade EF-137	8.8 (28 10.5)	c14.35 (47 1) including cannon	22.5 (241.9)	6,405 (14,120)	2 x Jumo 004C 11.75 (2,645)	1,000 (622) at S/L, 990 (615) at 5,000 (16,404)	1 x 37mm + 2 x 23mm cannon
Lavochkin '160' (flown)	8.95 (29 4)	9.12 (29 11)	15.9 (171.0)	3,837 (8,459)	1 x RD-10YuF 11.5 (2,580)	960 (597) at S/L, 970 (603) a t5,000 (16,404)	2 x 37mm cannon
Lavochkin '168' (flown)	9.5 (31 2)	10.56 (34 7)	18.08 (194.6)	4,412 (9,727)	1 x Nene 22.2 (5,000)	1,000 (621) at S/L, 1,084 (674) at 2,500 (8,202)	1 x 37mm + 2 x 23mm cannon
Lavochkin '174D' Prototype (flown)	8.90 (29 2.5)	9.51 (31 2.5)	16.16 (173.8)	3,708 (8,175)	1 x Derwent 5 15.6 (3,500)	900 (559) at S/L, 1,040 (646) at 3,000 (9,843)	3 x 23mm cannon
Lavochkin '176' (flown)	8.59 (28 2) 26.5 (5,952)	10.97 (36 0)	18.25 (196.2)	4,631 (10,209)	1 x VK-1	1,105 (687) at 7,000 (22,966)	1 x 37mm + 2 x 23mm cannon
Mikoyan I-310 1st Prototype (flown)	10.08 (33 1)	10.2 (33 5.5)	20.6 (221.5)	4,840 (10,670)	1 x Nene I 20.0 (4,500)	905 (562) at S/L, 1,028 (639) at 5,000 (16,404)	1 x 37mm + 2 x 23mm cannon
Mikoyan MiG-15 Early Production (flown)	10.08 (33 1)	10.1 (33 2)	20.6 (221.5)	4,915 (10,835)	1 x RD-45F 22.2 (5,000)	1,052 (654) at S/L, 1,020 (634) at 5,000 (16,404)	1 x 37mm + 2 x 23mm cannon
Mikoyan MiG-17 Early Production (flown)	9.628 (31 7)	11.09 (36 4.5)	22.6 (243.0)	6,072 (13,386)	1 x VK-1 26.5 (5,952)	1,070 (665) at 5,000 (16,404)	1 x 37mm + 2 x 23mm cannon Later marks had 4 x RS-1-U AAMs
Mikoyan I-340 (flown)	9.628 (31 7)	11.264 (36 11.5)	22.6 (243.0)	5,210 (11,483)	2 x AM-5A 21.1 (4,740)	1,193 (741) at 5,000 (16,404)	1 x 37mm + ? x 23mm cannon
Sukhoi Su-17	9.6 (31 6)	15.253 (50 1)	27.5 (295.7)	7,390 (16,292) research, 7,890 (17,394) fighter	2 x TR-3 39.2 (8,818) dry, 45.1 (10,141) reheat	Research aircraft: 1,252 (778) at S/L, 1,152 (716) at 10,000 (32,808)	2 x 37mm cannon (not fitted to research prototype)
Yakovlev Yak-30 (flown)	8.65 (28 4.5)	8.58 (28 2)	15.1 (162.4)	3,330 (7,341)	1 x RD-500 15.6 (3,500)	1,010 (628) at 5,500 (18,045)	3 x 23mm cannon
Yakovlev Yak-40	5.05 (16 7)	7.5 (24 7)	?	1,800 (3,958)	2 x axial jets 8.3 (1,874)	?	2 cannon
Yakovlev Yak-M	7.7 (25 3)	10.25 (33 7.5)	?	3,000 (6,614) normal 3,450 (7,606) overload	1 x AM-5	1,158 (720) at 1,000 (3,281), 1,131 (703) at 5,000 (16,404)	1 x 37mm + 2 x 23mm cannon

Chapter Three

Supersonic Generation

By the early 1950s Soviet fighter designers were considering and planning new aircraft that were to be capable of exceeding Mach 1 on the level with ease. Like other teams around the world they were helped in this by the development of more powerful jet engines and the adoption of advanced aerodynamic wing shapes. Sukhoi and Mikoyan were to take the delta wing very seriously, the latter producing one of the best fighters of all time in the form of the MiG-21. By the mid-1950s the main players in the fighter design field were Mikoyan, Sukhoi and Yakovlev, although the Lavochkin OKB had still to make its final contributions (described in Chapters Four and Five). This chapter, however, also includes the only supersonic fighter study to be made by the Antonov Design Bureau.

Arkhip Lyul'ka was to produce a series of outstanding engines for Soviet military aircraft. His AL-5 was referred to in the previous chapter but it was used only by certain prototypes, so the first Lyul'ka engine to be put into mass-production was the more advanced AL-7, the initial example being built in March 1953. Eventually different versions of this outstanding engine, with ever-increasing thrust ratings, were produced in large numbers for both fighter and bomber types. A concurrent engine was the much less powerful AM-5 from the Aleksandr Mikulin Design Bureau that was used by the Mikoyan I-340 described in Chapter Two. This was a single-shaft engine that was very slim in diameter and could thus be used in multiples within a fighter airframe, but it lacked thrust and a developed version was subsequently produced called the AM-9 (which was later renamed the RD-9B). Various marks of this unit were employed in versions of the Mikoyan MiG-19.

Unorthodox Supersonic Designs

To begin this study of the Soviet Union's first supersonic fighters there is a pair of designs that were, in appearance, quite similar. Neither flew (although the second almost got there) and they are followed by some very advanced proposals from Boris Cheranovskii.

Antonov Supersonic Fighter

In December 1952 the Design Bureau of Oleg K Antonov had another go at fighter design when it began studies for this supersonic interceptor project. The Central State Aerodynamic and Hydrodynamic Institute at Zhukovskiy (TsAGI) had supplied research data and information relating to the theories of low-aspect-ratio wings and the resulting preliminary project was based on this material. The drawing showed a tailed, low delta-winged aeroplane of small span, a nose intake and one AL-7F engine, plus three 30mm cannon mounted internally. Preliminary estimates indicated that the aircraft would achieve a top speed in the region of 1,800km/h to 1,900km/h (1,119mph to 1,181mph) at a height of 10,000m (32,808ft). It would reach 15,000m (49,213ft) altitude in three minutes, have a ceiling between 19,000m and 20,000m (62,336ft and 65,617ft) and an endurance of two and a half hours on its internal fuel. TsAGI looked on the project with favour but the Ministry of Aircraft Industry (MAP) refused to put it into its experimental aircraft programme for 1953, so the design stayed on the drawing board.

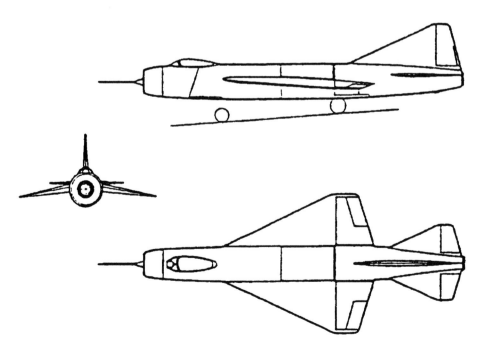

Engine designer Aleksandr Mikulin.

Antonov Supersonic Fighter (12.52).

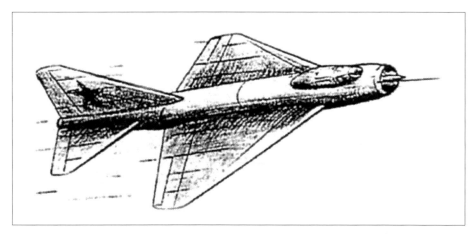

Rough sketch of the Antonov Supersonic Fighter project.

Model of Antonov's Supersonic Fighter. George Cox

15.6kN (3,505 lb) thrust RD-500 Derwent copy was chosen. Accordingly the design performance figures were much more modest and included a take-off weight of 2,470kg (5,445 lb), including 500kg (1,102 lb) of fuel, and a top speed of 1,100km/h (684mph). Project leader was Leonid L Selyakov and the aircraft was called the Yak-1000, it had a zero-track bicycle undercarriage and carried no weapons. During November and December 1950 a full-size wooden mock-up was tested in TsAGI's wind tunnels.

The decree covering the Yak-1000 programme stated that the prototype was to be delivered for flight test on 1st January 1951, but this deadline was missed and the only example to be built was not completed until 27th February. Taxi tests were under way from 2nd March and at speeds below 100km/h (62mph) the pilot reported good directional stability, easy turns and instant acceleration. However, the next day the Yak-1000 was taken up to 250km/h (155mph) and a strong crosswind from the starboard side, registering 5m/sec (16.4ft/sec), made the aircraft lean onto its port outrigger leg. Opposite aileron failed to counter the problem effectively and the aircraft was forced off the runway. A week later, on 10th March, it ran off the runway again, this time in a slight crosswind, and so the programme had to be suspended. A list of possible corrective modifications, actually completed within four days, included alterations to the rear fuselage, undercarriage and control systems, plus a change to the CofG position. One proposal showed the vertical tail replaced by two end-plate fins on the tips of the horizontal tail. A model of the new configuration was tested at TsAGI, but to no avail – all work on the Yak-1000 project was cancelled officially in October 1951 without any flight-testing taking place. Most considered this to be a sensible move although one or two of Yakovlev's design team disagreed.

Problems with the undercarriage at take-off were in fact only part of the trouble. It was also found that the new wing posed serious stability problems in flight that could not readily be tackled at this stage. In truth the Yak-1000, or rather its configuration, was ahead of its time and the current state of the art was not ready to handle the stability and

Yakovlev Yak-1000

Yakovlev wanted to produce a tactical fighter that would become a worthy rival to the aircraft coming out of the Mikoyan OKB. This wish was backed up by a government directive issued in June 1950 that covered a fighter to be powered by the new Lyul'ka TR-3 (AL-5) by-pass turbojet, then under development. The layout embraced some strong recommendations from a TsAGI aerodynamics expert, Pyotr P Krasil'shchikov, to make use of German research on the so-called rhomboid-shaped wing. In fact the planform of the wing on the resultant aircraft was more like a cropped delta with a sharply swept leading edge and a moderate negative sweep (taper) on the trailing edge. The wing itself was very thin with a t/c ratio of 3.4% at the root and a maximum elsewhere of 4.5%.

The specification within the directive requested a performance that was unusually high for its time, including a top speed of Mach 1.7. When the Yakovlev team started work it was faced with the problem that the AL-5 engine was still unavailable and this prompted a decision to build a 'technology demonstrator' research aircraft with the rhomboid wing to test its aerodynamics and airfield performance. It would be powered by an available production engine and a single

Wind tunnel model of the Yak-1000.

Soviet Secret Projects: Fighters Since 1945

Left and above: **The sole Yakovlev Yak-1000 research aircraft was destined never to fly.**

Below: **Two views of what is believed to be a model of the fighter project that was based on the Yak-1000.**

control weaknesses that were inherent with this design. It was not until further research had become available from TsAGI that the USSR's fighter designers could approach the problem of building aircraft with delta wings with more confidence – the results of that are described shortly. The original Lyul'ka TR-3-powered fighter project was also terminated in October 1951.

Yakovlev Yak-1000M/Myasishchev M-33

One of the proposed solutions for the Yak-1000's problems was a project called the Yak-1000M. A drawing showing the general layout of the new version was signed by Leonid Selyakov on 17th March 1951 and it revealed the following changes – a wing set lower on the fuselage and given marked anhedral and different incidence, a re-designed vertical tail with more area and sweepback, and the tailplane moved from its position on the fin down to the rear fuselage. It appears that this project did not receive an approval to proceed from the Yakovlev OKB.

During the same month, however, the government issued a directive establishing a special design bureau under the control of V M Myasishchev specifically to develop a strategic bomber (the future Myasishchev M-4 *Bison*). On 31st March Selyakov and numerous other engineers were transferred from Yakovlev to the newly-formed OKB and he took the Yak-1000M design with him. Myasishchev redesignated it M-33 and there were plans to use this aircraft for research into the behaviour of low-aspect-ratio delta wings at transonic speeds. The Yak-1000M had one 17.6kN (3,970 lb) thrust TRD-5 engine and the performance estimates suggested that it would be supersonic at height; the estimated service ceiling was 15,000m (49,213ft) and there appears to have been no armament. It was not built.

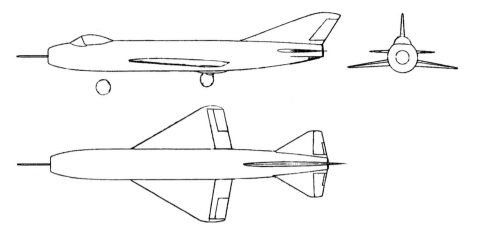

Cheranovskiy Designs

Cheranovskiy Jet Fighter

Boris I Cheranovskiy was an innovative designer who produced a series of glider and powered aircraft prototypes and projects from the 1920s into the 1950s, almost all of them using some form of flying wing and often featuring other elements of design that were well ahead of their time. Many of his proposals were outstanding but Cheranovskiy always seems to have suffered from a lack of available money to see his aircraft through. In 1944 he drew an unnumbered single-seat delta wing jet fighter project with two widely-spaced engines, two cannon in the nose and no vertical tailplane, and this remarkable project would not have looked out of place amongst today's 'stealth' programmes. It was known either as 'the flying wing' or 'the jet flying wing' and had its ailerons protruding outside the wing outline. Engine type is unknown but it offered 4.9kN (1,100 lb) of thrust, although whether that came from one or both units is also unknown. With this powerplant the estimated maximum rate of climb at sea level was 24m/sec (79ft/sec), time to 1,000m (3,280ft) 42 seconds and range when flying at maximum speed 426km (265 miles).

A report examining the jet fighter was completed by V S Pyshnov, Major General for the Air Force's Aviation Engineering Service, on 27th December 1944. He declared that 'this proposal deserves due attention, as the high strength and rigidity combined with the minimum weight is of paramount importance to a high-speed jet. The proposed solid structure suits this objective best of all. It should be noted that at a near-sonic speed, which can be achieved in descending flight, the standard tail unit turned out to be inefficient. It is hard to say whether Cheranovskiy's design will be better, but I think that the control surface protruding out of the wing will be more efficient'. Cheranovskiy had estimated that his project would have a top speed of 800km/h (497mph) at ground level and Pyshnov noted that, given the thrust of 4.9kN (1,100 lb), a speed of 750km/h (466mph) was 'quite realistic'. He concluded by saying that the design 'will naturally require thorough research and development work, but that holds true for all jet aircraft being built. B I Cheranovskiy has abundant experience in design work and inventions, as well as

Above: **Yakovlev Yak-1000M/Myasishchev M-33 (17.3.51).**

Left: **Cheranovskiy's unnumbered jet fighter project of 1944.** Russian Aviation Research Trust

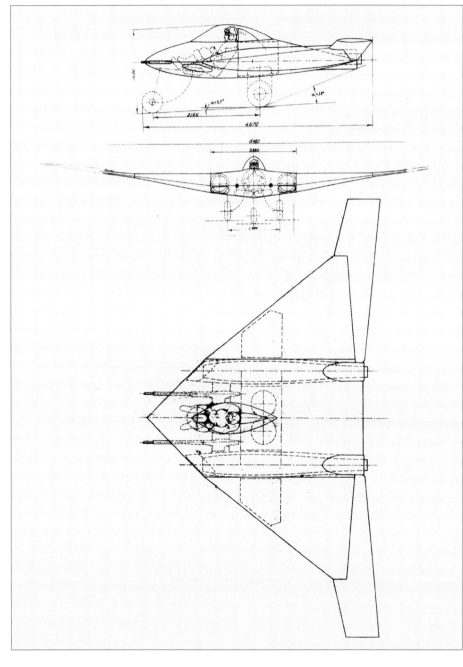

Soviet Secret Projects: Fighters Since 1945

Model of the Cheranovskiy BICh-24.

Cheranovskiy BICh-26 (c1948).

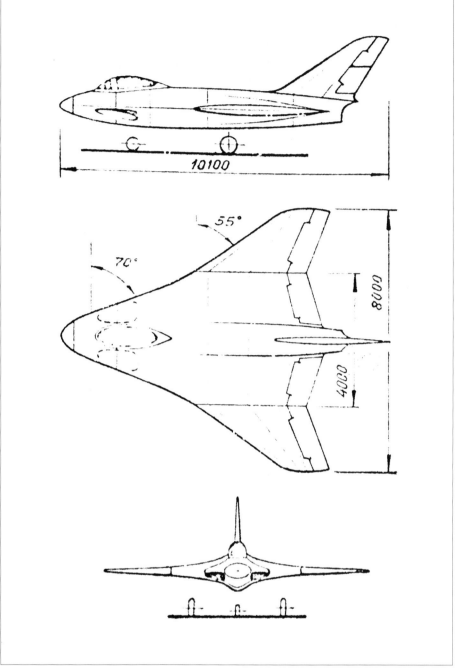

energy, persistence and initiative. Given a required team of workers, he is capable of bringing the project to the result desired'.

There is no information regarding the fate of the project but no hardware appears to have been produced. Nevertheless Cheranovskiy continued his research and a similar flying wing layout was to appear later in two experimental glider designs: the BICh-22 (Che-22) and BICh-23 (Che-23). The former was drawn during 1947/48 and had wingtip-mounted fins; it was intended to examine the slow-speed characteristics of 'the jet flying wing' and it made its maiden flight on 3rd June 1948. In fact Cheranovskiy's reputation had been enhanced during the war and after the conflict was over he was able to build up a small OKB design team to help him. In 1947 this OKB was based within the Moscow Aviation Institute (MAI) where some superb facilities were available to assess Cheranovskiy's most advanced designs. A small production batch of BICh-22s was eventually built and the follow-on two-seat BICh-23 development, with greater leading edge sweep, made its maiden flight on 7th February 1949.

Cheranovskiy BICh-24

This project was intended to investigate the tailless delta configuration. With the advent of the jet age Cheranovskiy recognised that he should think in terms of a much lower aspect ratio and he followed his 1944 project with the graceful BICh-24 jet fighter. He hoped to demonstrate a prototype in the Tushino 'parade' of 1949 but it appears that wind tunnel testing was as far as the project progressed. The tunnel model looked very similar to the BICh-26 configuration described below but few details and no data have been found for the BICh-24, which in some documents was called the Che-24.

Cheranovskiy BICh-25 and BICh-26

The flying wing research continued well into the 1950s and included various experiments

with model wings of different shapes. References have been found to another project, the BICh-25 (Che-25) jet fighter design produced in June 1948, but little information has been traced although it apparently had a variable-sweep wing with an outboard wing pivot position. However, it was followed by a beautiful experimental tailless delta jet fighter project with blended leading edges and a large fin called the BICh-26 (or Che-26).

The BICh-26's single fin carried both upper and lower rudders, there were inboard elevators and outboard ailerons on the wing trailing edge, flush inlets in the underside of the

flattened forward fuselage level with the cockpit, and a tricycle undercarriage. Power came from a single Mikulin AM-5 turbojet and the estimated performance figures included a service ceiling of approximately 22,000m (72,178ft). In appearance the ambitious BICh-26 looks to have been a very advanced and outstanding project. It was also intended to be very supersonic and capable of reaching Mach 1.7 at 7,000m (22,966ft), which is why the whole section on Cheranovskiy has been placed here. Over a period of several years much of the BICh-26's calculations and tunnel testing was completed and some

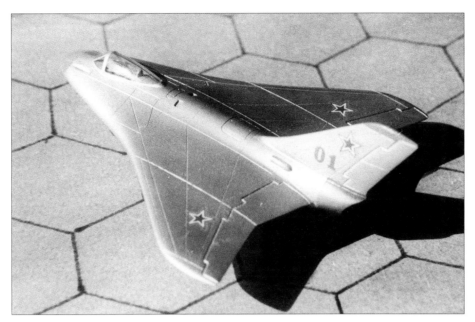

Left: **Model of the BICh-26.**

Below left: **Artist's impression of the beautiful BICh-26.**

demonstration models were built. Cheranovskiy died in 1960 and sadly never saw any of his jet fighter designs turned into hardware.

–

While work was proceeding on these unusual fighters, Mikoyan, Sukhoi and, to a lesser extent, Yakovlev had begun programmes that were to produce a lot of prototypes, and also lead to some highly successful supersonic aeroplanes.

Mikoyan MiG-19 Background

Mikoyan I-350

On 10th June 1950 the Mikoyan OKB was ordered by a Council of Ministers decree to produce a new fighter capable of full supersonic performance, which was to be powered by Lyul'ka's TR-3 (AL-5) engine. Leadership of the new programme, designated I-350 and '*Izdeliye* M', was given to A V Minayev and the aircraft retained the nose intake of Mikoyan's earlier jets together with the tail placed halfway up the fin. However, coupled with these features was a new wing swept 60° at the leading edge, which also had four boundary layer fences on each side to prevent airflow outwards towards the tips, plus a longer and sleeker fuselage. This wing was a big achievement for 1950 because it used a suitably low t/c ratio to allow supersonic flight, but had sufficient stiffness to ensure that powerful ailerons could be fitted which would not give a reverse of lateral control when flying supersonically. The nose was similar to the MiG-17P variant with an RP-1 Izumrud (Emerald) radar fixed on the intake lip.

There was to be just the one I-350/M prototype, which made it into the air for the first time on 16th June 1951. Its flight test programme was to be short, just five flights in all, because several sorties were terminated by problems with the engine which showed tendencies to flame out; consequently a full set of performance figures for the I-350 was never obtained. However, the type's service ceiling was 16,600m (54,462ft) and time to 10,000m (32,808ft) 2.6 minutes. The M-2 project was an unbuilt I-350 development with a

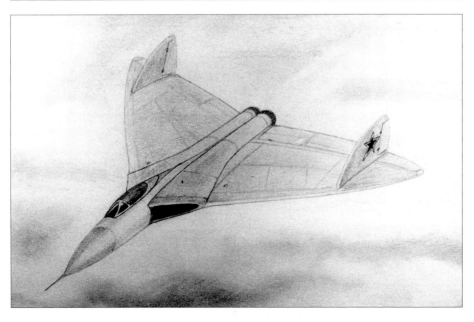

Impression of another unknown Cheranovskiy jet fighter design. This appears to be rather larger than the BICh-26 and has twin engines and outer wing fins.

Soviet Secret Projects: Fighters Since 1945

Korshun radar in a radome mounted in the centre of the inlet, and proposals were also made for a trainer variant called the MT. The weakness of the I-350 was its underdeveloped engine, which was also fitted in the unflown Sukhoi Su-17 (Chapter Two), but the potential was there; in effect it was to have been a 'super-fast' MiG-17. In due course a reheated TR-3F was to have been fitted. However, at this stage the Mikulin AM-5 power unit had reached a better level of reliability and two of these were used in a parallel Mikoyan project, the I-360.

Mikoyan I-360

The I-340 project detailed in Chapter 2 used two Mikulin AM-5 engines. This all-new escort fighter project was known as the SM-2 (the I-340/SM becoming the SM-1) and was also designed to use two AM-5A power units. The I-360 was given the go-ahead by a directive issued on 10th August 1951, although its preliminary design had already been under way for some time. The design was completed in December with wings of slightly less sweep on the leading edge (57°) than on the I-350 and, although shorter than that aircraft, it had a fuselage 1.6m (5ft 3in) longer than the MiG-17. Two machines were built, plus a static test airframe, and the first of these made its maiden flight on 24th May 1952. As built, the I-360 had its horizontal tail placed near to the top of the fin, but this gave problems with spin recovery and was eventually moved to the base of the fin. The performance data was encouraging and, after a thorough evaluation, the type was recommended for production, pending the elimination of several weaknesses. The OKB, however, went on to design and build a new fighter based on I-360.

Mikoyan SM-9 and MiG-19

Despite showing good performance, the AM-5's lack of power was still a problem. However, in 1953 the follow-on AM-9F appeared which was a straightforward development with the same dimensions but having an extra stage of blading in the compressor, a new combustion chamber, reheat and a convergent-divergent nozzle. This engine fitted within the basic I-360 airframe seemed an ideal solution and another Council of Ministers directive was issued on

Top: **Mikoyan I-350/'M'.**

Centre: **The first Mikoyan I-360/SM-2 prototype as originally flown with a high tailplane position. Note the guns housed in the wing roots.**

Bottom: **The first I-360 after the tail had been moved to the lower position.**

Mikoyan SM-9 (MiG-19) first prototype.

Mikoyan SM-7 (MiG-19P) second prototype.

Mikoyan I-370/I-1 prototype.

15th August 1953 requesting two versions of this modified design, a tactical fighter and an interceptor. The first SM-9 prototype was rolled out on 21st December 1953, in appearance near-identical to the I-360/SM-2, and made its maiden flight on 5th January 1954. Two more prototypes followed and the first production machines entered service as the MiG-19; the ASCC gave it the codename *Farmer*. This was the first production Soviet jet fighter designed from scratch to be supersonic and it used a version of the wing first tried on the I-350 above. It proved to be another highly successful service aircraft and 1,884 were built in the Soviet Union alone.

The interceptor outlined by the August 1953 directive was satisfied by another version of the MiG-19, which began life in prototype form as the SM-7. A slightly longer and modified forward fuselage housed an RP-1 Izumrud radar, a wider cockpit and redesigned intake ducting. This left no room for nose-mounted guns and so just two 23mm cannon were carried, one in each wing. The first flight was made on 24th August 1954 and the type entered service as the MiG-19P (P = *perekhvahtchik* or interceptor); in the West it became *Farmer-B*.

Mikoyan I-370

The August 1953 'MiG-19' directive had in actual fact requested two versions of the tactical fighter, with the second an 'insurance' type powered by a single reheated centrifugal VK-7 engine from the Klimov Design Bureau. This new powerplant, specially designed for the resulting I-370 prototype (which was also known as the I-1), proved to be the world's most powerful jet engine of the centrifugal type, but one of the least successful.

As noted in Chapter One the Klimov VK-1 had been designed in the USSR as an enlarged copy of the British Nene. It was to spawn a family of engines but none of these was to match the original production type's success. The first was the VK-2 turboprop of 1947, which did not go into series production. Then came the VK-5 and VK-7 developments that were also only produced as prototypes. The VK-5 was a further development of the VK-1A with the maximum thrust increased to 30.4kN (6,835 lb), although the structure was still basically the same as the VK-1. A VK-5F version with afterburning was also produced that gave a maximum 51.4kN (11,575 lb) with reheat. Compared to the VK-1F, this unit sported a new afterburner with a variable exhaust nozzle and it passed its State bench test programme in 1953; it was eventually flight-tested on the MiG-17R reconnaissance aircraft.

The VK-7 was designed in 1952 as a development of the VK-5. It had a layout based on the VK-1A but received a new two-stage compressor, impellers of complex shape, modified combustion chambers, reheat and many other minor changes. Final thrust ratings, as the VK-7F, were 41.1kN (9,260 lb) dry and 61.4kN (13,825 lb) with afterburner and it became the last Soviet jet engine to be designed with a centrifugal compressor – all future types would have axials. In fact, in between this pair of centrifugal designs, Klimov had produced the VK-3 axial-flow turbojet with a ten-stage compressor, circular combustion chamber, three-stage turbine and afterburner. Its development began in 1949 and the engine was tested in 1952, giving a thrust of 56.1kN (12,630 lb) dry or 82.7kN (18,605 lb) with reheat. Improvements to the VK-3 continued until 1955 and in 1953 Mikoyan began the design of the I-3 (I-380/I-420) fighter (Chapter Five) with this powerplant in mind. However, the VK-3 could not withstand the competition pre-

sented by the Lyul'ka AL-7 and so did not enter production.

Returning to the I-370, this project was originally submitted for its official review with a T-tail, rather like the I-360, and as such the design was approved. However, the spin recovery problems inherent with this arrangement were discovered during the earlier aircraft's flight test programme and this revelation prompted an immediate change on the I-370 to a conventional tailplane. The aircraft had identical wings to the I-360 and the forward fuselage was also similar but slightly longer with a smaller intake to satisfy the VK-7; the rear fuselage, however, had to be redesigned to accommodate the new engine. In due course problems were experienced in getting the VK-7 to provide its specified thrust that, following the I-370 prototype's completion on 20th May 1953, resulted in the aircraft sitting on the ground for a long period waiting for its powerplant.

Such substantial delays were to prove fatal because at the end of 1954 a new decree was issued cancelling the VK-7 project altogether, although the prototype I-370 was eventually

allowed to make its long-overdue first flight on 16th February 1955. On test it showed a similar top speed to the MiG-19 but an inferior rate of climb. In the spring of 1956 the sole prototype was passed to the LII Flight Test Research Institute to be used in a research and development project. Having initially been fitted with an early VK-7 giving 51.3kN (11,540 lb) of thrust in reheat, the prototype eventually received a more powerful VK-7F. The construction of a second prototype also reached an advanced stage. This machine was redesignated I-2 and it was fitted with a new wing of greater sweepback, although the span was unchanged. However, the severe delays experienced by the I-1 prevented the completion of this follow-on programme and work on the I-2 prototype was terminated when it was about 93% complete.

Mikoyan SM-12

Following behind the MiG-19 came a series of prototypes that were intended to examine various modifications to the air intake/shock cone combination. The MiG-19S (*Farmer-C*) variant of the basic fighter was the original

Artist's impression of the Mikoyan I-2.

The third Mikoyan SM-12 armed with underwing rocket pods.

MiG-19 fitted with an all-moving tailplane instead of the original fixed version with elevators. On 6th September and 12th December 1956 MAP introduced two decrees that requested improvements to the MiG-19S's aerodynamics. In addition uprated engines were to be fitted and, later on, a new version of the RD-9 called the R3-26 was also introduced, developed by V N Sorokin who was the leader of a division of the Mikulin OKB based in Ufa. This resulted in the SM-12 series of four prototypes, all converted from production MiG-19S airframes, and the first of

them flew on 19th April 1957. The most visible difference was a longer nose with a sharp-lipped MiG-21F-type intake, plus a movable shock cone centrebody (the MiG-21 is described shortly). There were no other major changes to the structure, although the extra weight limited the gun armament to two NR-30 cannon in the wings.

The increases in speed, ceiling and overall performance achieved with this upgraded aircraft were most marked and well into the supersonic regime, but the higher fuel consumption reduced the type's range; the

SM-12's service ceiling was 17,500m (57,415ft). There were other problems, particularly with the operation of the engine, but from a combat point of view the SM-12 was considered to be almost equal to the Ye-2A and Ye-5 (below) and could be in production much earlier. The test pilots from the Soviet Air Force Scientific and Research Institute (NII VVS) recommended that the type should replace the MiG-19S on the production line, but the Ye-6 (MiG-21 prototype) was clearly the more promising aircraft and in June 1958 the SM-12 programme was terminated on official orders; however, the third and fourth SM-12s were kept in the air to serve as test aircraft for new air-to-air missiles. After this six more prototypes of another version followed, designated SM-12PM, which had a larger sharp-lipped intake and bigger shock cone. The first of these flew on 27th May 1958 and they served as development and research aeroplanes for various systems and pieces of equipment.

Mikoyan MiG-21 'Family'

Ye-1, Ye-2, Ye-50, Ye-2A, Ye-4 and Ye-5

During the 1950s the rate of progress in the development of supersonic airframes was such that the advanced MiG-19/SM-12 developments got rather left behind; in fact the SM-12s were probably looked on as no more than insurance designs against the failure of the Design Bureau's latest projects. These comprised a series of prototypes that basically used the same fuselage but had alternative engine and wing options, the idea being to find the most suitable combination to put into production as a lightweight fighter with a top speed well above Mach 1. In due course the resulting MiG-21 would be capable of Mach 2.

The first of the prototypes was designated Ye-1. It had its wing swept 57° on the leading edge and was intended to use a new engine from the Mikulin OKB called the AM-11, a power unit designed specifically for fighters like the Ye-1 by a team led by Sergei K Tumanskiy. In 1955 Mikulin lost the control of his Design Bureau in a political move and the organisation was renamed after Tumanskiy, who became its new leader. In 1956 the new engine was redesignated RD-11, but it would not be available in time for the Ye-1 so a single, less powerful AM-9B (MiG-19) power unit had to be substituted, and the prototype was redesignated Ye-2 to reflect this. The Ye-2 first flew on 14th February 1955.

As soon as it was realised that the AM-11 would be late, and to address the shortfall in thrust created by having the AM-9B, Mikoyan designed a parallel version of the Ye-1/Ye-2 fitted with an auxiliary Dushkin S-155 rocket motor mounted in the root of the fin above the jetpipe. Three prototypes were built, which were designated Ye-50, and the first of these flew on 9th January 1956. This type is described more thoroughly in Chapter Seven but, after the Dushkin design team had closed on Ministry orders, the rocket programme and follow-on Ye-50A project were abandoned. Official policy had switched to jets only.

The next prototype in the main series was the Ye-2A, which had the AM-11 in place. The Ye-2 and Ye-2A were complemented by the Ye-4 and Ye-5, which received the AM-9B and AM-11 respectively, plus a delta wing with the leading edge also set at 57°. In other respects all of these aircraft were practically identical and featured a nose intake with a conical centre body and an all-moving 'slab' tailplane.

The second to fly was the Ye-4 on 16th June 1955, and when the AM-11 finally became available the Ye-2A and Ye-5 made their first flights on 17th February and 9th January 1956 respectively. The Ministry of Aircraft Industry felt certain that one of Mikoyan's two designs would be ordered into production, so in late 1955 it allocated the service designations MiG-21 and MiG-23 to the delta Ye-5 and swept wing Ye-2A respectively. A pre-production batch of five aeroplanes was built for each AM-11 variant and intensive flight-testing confirmed that the delta offered a higher rate of roll, a better turn radius, lighter structure and more fuel, and a very slight improvement in supersonic performance. Initially however, the Ye-4 did show a shortfall in top speed that prompted some modifications to the wing fittings.

With flight experience, the Design Bureau was able to make quite a number of improvements and modifications to the basic delta design; the more mature R-11F-300 engine was also introduced. Consequently a new series of delta prototypes was built under the designation Ye-6, the first flying on 20th May 1958, and in due course the design entered production and service as the MiG-21. This fighter was capable of reaching twice the speed of sound and proved to be another incredibly successful Mikoyan design – many examples are still in service today. However, prototypes of both wing shapes were flown in front of Western observers at the Tushino air display of 24th June 1956, which resulted in the codenames *Faceplate* and *Fishbed* being allocated to the swept- and delta-winged aircraft respectively – in fact it was expected that the former would be the main service aeroplane. The MiG-23 designation was later re-used for the variable-sweep fighter described in Chapter Eight.

Above right: **Mikoyan Ye-5.**

Right: **Mikoyan Ye-6, the first prototype MiG-21.**

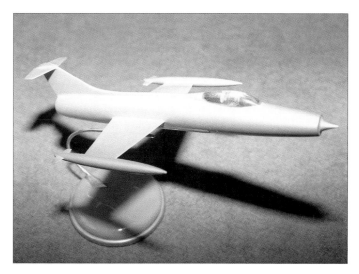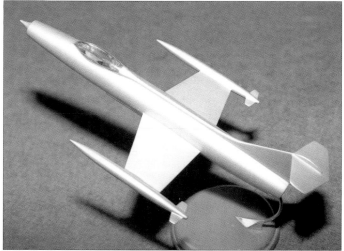

Two views of the proposal which showed a MiG-21-style fuselage mated with wings and tailplane that were clearly inspired by the F-104 Starfighter.

One line of MiG-21 developments introduced an all-weather interceptor capability. In July 1958 a resolution was issued by the Council of Ministers requesting Mikoyan to test the K-13 air-to-air missile and TsD-30 radar aboard the MiG-21 airframe. The resulting Ye-7 prototype first flew on 10th August 1959 with a larger central nose cone for the radar and one missile beneath each wing. The project led to production versions including the MiG-21PF, MiG-21PFS and MiG-21PFM with a larger dorsal spine. These were produced in the early 1960s and became a mainstay in the Soviet Air Force fighter fleet.

Mikoyan 'MiG-21 Variant' and '7-31'
Material has come to light on two alternative 'versions' of the MiG-21. The first was based on the original Ye-series designs but was clearly inspired by the American Lockheed

F-104 Starfighter. It has the basic Ye-1/Ye-6-type fuselage and swept fin, but has lost the swept tail and swept or delta wing for an F-104-style near straight wing and a T-tail. In fact the wing shape, the shape of the tailplane and its position on the fin, and the long wingtip tanks, are remarkably similar to the XF-104A first flown in February 1956. One assumes that that was also the approximate date for the project. The design was apparently based on suggestions made by the Bureau's designers and appears to have been an official study.

The date of the projected '7-31' development is unknown but it appears to have originated in the 1960s. Compared to early standard MiG-21s the '7-31' had a different wing with small leading edge root extensions, a thicker dorsal spine and it carried a single AAM on each inner wing pylon and two on

each middle pylon; a streamlined drop tank was also carried under the outer wings. In 1976 this project was revised as the '7-33' with a single RD-33 engine as used by the new MiG-29. No data is available for any of these proposals.

Sukhoi Su-7, Su-9 and Su-11 Background

Soviet copy of North American F-86 Sabre
After the death of Soviet leader Joseph Stalin in 1953, Pavel Sukhoi returned to favour. He replaced V V Kondrat'yev as the head of OKB-1 near Moscow, and in early 1954 the Bureau was re-numbered OKB-51. During October 1946 this OKB had been formed with German engineers who had been taken to Russia from the Junkers company at the end of the war. It had previously operated under the leadership of Brunolf Baade and later Semyon Alekseyev. In his new role Sukhoi also acquired the job of copying (or rather reverse-engineering) the American F-86 Sabre swept-wing jet fighter, a project which had already been awarded to OKB-1 following a decision made in October 1952 to copy the aircraft. Two examples of this excellent fighter were acquired from Korea during the conflict of 1950 to 1953 and consideration was given to producing a version with Soviet engines. However, Sukhoi was unhappy with the idea and pushed the new Ministry of Aircraft Industry (MAP – formerly NKAP, the People's Commissariat for Aircraft Industry) to do an all-new design of his own. Permission was granted and the resulting S-1 replaced the

Mikoyan '7-31' project.

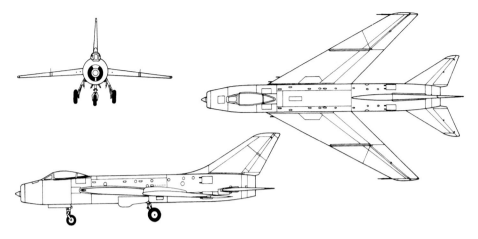

'Sabre copy', which was then abandoned. The F-86 had been studied in great detail and its acquisition had put many other new technical advances into the Soviet Union's hands, including the fighter's APG-30 radar.

Sukhoi S-1, Su-7 and S-3

The experience gained by the OKB's design staff from working on the F-86 proved beneficial to the new proposals. In addition N G Zyrin became Sukhoi's deputy for airframe design and he was to play a key role in the development of the forthcoming family of jet fighters. Like Mikoyan above, a big decision had to be made regarding what type of wing was to be used. During the early 1950s there was a great deal of discussion within the USSR's aircraft industry about the benefits of delta wings over swept, particularly within the TsAGI research institute – deltas were great for high speed but fell down a little when it came to manoeuvrability. In fact, prior to his replacement as the OKB's leader, Kondrat'yev and the OKB-1 team had considered the delta to be the way forward.

As a result prototypes of Sukhoi's swept- and delta-winged projects were both built, which mirrored the work done at Mikoyan. In fact the series that follows became something of a rival to the 'MiG-21' prototypes described above, but Sukhoi's aircraft were bigger and heavier overall. In the end there was room to put both Mikoyan's and Sukhoi's aircraft into production, the latter doing a great deal to establish Sukhoi as a major player in fighter aircraft design. In the end the production MiG-21 had a delta wing and, despite the success of the Su-7 swept wing fighter-bomber, Sukhoi also returned to the delta for his follow-on designs, keeping faith with it well into the 1960s.

The S-1 tactical fighter proposal (S for *strelovidnoye krylo* or swept wings) was completed in November 1953 in response to a Council of Ministers directive issued three months earlier. This directive was actually the go-ahead for Sukhoi's all-new project that had been requested ahead of the F-86 and it covered two types, a tactical fighter and an interceptor. The performance requirements included a maximum speed of at least 1,800km/h (1,119mph) for the tactical fighter, or 1,900km/h to 1,950km/h (1,181mph to 1,212mph) for the interceptor. As part of the

research the S-1 was joined by three other designs, the swept wing S-3 interceptor of mid-1954 which was intended to fill the second part of the requirement, and then the T-1 tactical fighter and T-3 interceptor of October 1954. The second pair had delta wings (T= *treoogol'noye krylo* or delta wing) but all four shared a roughly identical fuselage with a nose intake and a single Lyul'ka AL-7F powerplant.

Mock-ups for all four were approved but the S-3 was dropped by a MAP decision made before work on a prototype had started. The S-3's design had been approved by the Air Force on 11th December 1953 and the main differences from the sister S-1 concerned the armament and the Almaz interception and fire-control radar. In fact the S-3 mock-up was the S-1 refitted as the interceptor with changes to the forward fuselage as far as the cockpit. The intercept antenna was housed in a radome placed over the air inlet while inside the inlet itself was a cowling containing the target-aiming radar. The armament was unchanged except for a reduction to just two Nudel'-man/Rikhter NR-30 30mm cannon, rather than the S-1's three, mounted in the wing roots.

Once again, the Soviet Union had embarked on producing a series of competing prototypes. As built the S-1 differed a little from the original 1953 design, the principal change concerning the main undercarriage legs that were now housed in the inner wing rather than the middle fuselage. Detail design was completed during 1954 and the prototype made its maiden flight on 7th September 1955. Such was the Soviet Air Force's need for new supersonic aircraft that a decision was quickly made to put the S-1 into production as the Su-7. In April 1955 the S-1 had been flown to a speed of 2,170km/h (1,347mph), way over specification, and this proved a big help in getting the production order. The prototype did experience some problems, but some of these were simply due the fact that this was a faster aircraft than anything that had been flown by the armed services before. For example, knowledge of intake design for Mach 2 flight was based purely on theory, so experience had to be gained to find the best solution. In due course, as we shall see, the Sukhoi OKB was to examine air intakes more thoroughly than any other OKB, in fact probably more than any other fighter design team worldwide.

After being publicly displayed at the Tushino air show on 24th June 1956, the S-1/Su-7 received the ASCC codename *Fitter*, but its career solely as a fighter proved to be relatively short. A second prototype, called the S-2, embraced some revised requirements including a higher ceiling and the AL-7F-1 engine that gave more thrust. This aircraft was the real prototype for the Su-7 and flew in late August 1956. Production deliveries began in May 1958 but in practice the Su-7 was too heavy for its tactical fighter role, and also exhibited some limitations in its combat capability, so the relatively small number built only stayed with their frontline units until the mid-1960s. However, the effort in producing the Su-7 was not wasted because the aircraft proved to be an excellent fighter-bomber. The Soviet Union badly needed such a type and in 1956 Sukhoi began work on a new fighter-bomber variant designated S-22. The first prototype flew on 24th April 1959 and this

version entered service, in much greater numbers, as the Su-7B; the *Fitter* codename was retained. Su-7Bs served with numerous air forces and led to more advanced versions, including the Su-17 and Su-22 with variable sweep wings described in Chapter Eight.

Sukhoi T-1 and T-3

The advanced development projects for the two delta-winged variants were completed as a joint document and, after passing their mock-up inspections, the T-1 tactical fighter received the higher priority. A special research team had been assembled within TsAGI to examine the aerodynamics of all of these designs, its chief being one P P Krasil'shchikov. For the T-series this group recommended fitting a wing of 6% t/c ratio with a rounded leading edge plus all-moving slab tailplanes (the latter were also advocated for the Mikoyan Ye-2 and Ye-4 deltas above). In terms of their forward fuselages,

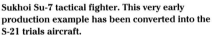

Sukhoi Su-7 tactical fighter. This very early production example has been converted into the S-21 trials aircraft.

the pair matched the S-1 and S-3 with the T-1 having a central shock cone and the T-3 the extra Almaz radomes; again the interceptor had just the two guns when T-1 had three. The T-1's detail design was ready in December 1954 but the T-3's was delayed by a switch from the original full monocoque structure to a semi-monocoque format with supporting longerons – T-3's detail design was finished in May 1955.

Another development to affect the T-3's schedule was the adoption of a new type of weapon system, the air-to-air missile (AAM), which was beginning to make its mark worldwide. This was done against a new Council of Ministers directive dated 30th December 1954 that specified either Toropov K-7L or Grooshin K-6V weapons. The former was a version of Toropov's first AAM design but tests with this missile were not a success and it was abandoned in the late 1950s. The rival medium-range K-6V was a high-altitude version of the first Grooshin AAM design and took so long to develop that, after test firings, it too was cancelled. Despite these difficulties, by the time the T-3's design was completed the delta interceptor had been given priority over the T-1 and the latter was cancelled before the end of the year. However, by then the T-1 prototype's construction was well advanced and

The Sukhoi T-3 prototype shows the Almaz interception and fire-control radome.

Soviet Secret Projects: Fighters Since 1945

Sukhoi decided to modify the airframe into the T-3 so that it could serve as the interceptor prototype, thereby removing the need to build an all-new airframe. The main alterations were the alternative nose and in the wing where one gun was taken out, but there were also changes to various items of equipment.

The T-3 made its first flight on 26th April 1956 and the machine was also displayed at the Tushino show. It thus received the code-name *Fishpot*, although this was later amended to *Fishpot-A* as described below. The flight test programme, spread over a year and a half, indicated that the basic layout was correct, but the T-3 was rather heavier than expected and the engine gave less thrust than predicted, so in some areas the performance did not match the specified limits. Results also showed that the fixed intake arrangement would prevent the fighter from achieving its specified top speed when using the AL-7F fitted with bleed valves on the fourth and fifth compressor stages. The bleed valves had to be introduced to help the engine's reliability but they also considerably reduced available thrust at flight speeds above Mach 1.6. Therefore the T-3 did not go into production since work was clearly needed to find a superior intake arrangement. T-3's service ceiling was 18,000m (59,055ft).

Sukhoi PT-7 and PT-8

The next step was a modified T-3 that was designed to carry AAMs from the start. It featured a new and quite gruesome intake but this offered a better arrangement for the avionics, including the Almaz-7, with radomes on both upper and lower lips. Prototype construction got going at the end of 1955 but there were problems acquiring the specified upgraded engine that delayed the first flight until the end of June 1957; as a result the trials timetable for testing the K-7L missile also had to be put back. The planned production version was designated PT-8 and three pre-production machines were ordered, although only the one was to be completed as such in February 1957. In April 1958 the Council of Ministers issued a new directive that introduced a new weapon system to Sukhoi's interceptor, necessitating yet more changes.

Sukhoi T-43 and Su-9

The catalyst for this particular set of developments was a blanket directive released in April 1956 to all fighter Design Bureaux that requested a higher ceiling. Hopefully some of the forthcoming aeroplanes could then deal with the latest American high-altitude reconnaissance types, such as the Lockheed U-2, which were breaking through Soviet airspace at an alarming rate. Soviet leader Nikita Khrushchev was extremely concerned about the situation and this kicked off an intensive effort to find a solution. In Sukhoi's case the S-1 and T-3's ceiling were pushed up to 21,000m (68,898ft), an increase of the order of 2,000m (6,562ft), which required the introduction of the new and more powerful AL-7F-1 engine; several items of equipment were also taken out to save weight. The slightly fatter power unit required some redesign to the rear fuselage and the necessary design work was finished in December – thirty production T-3s/PT-8s ordered for manufacture in 1957 would have to be modified.

The need for a higher ceiling and higher speed brought other changes. Principally the fixed intake arrangement on the PT-8 was clearly unsuitable and a more efficient variable intake with a central cone was adopted to create shock waves inside the inlet; these made the airflow slow down at a gradual rate, the optimum position depending on the aircraft's speed. In addition, rocket boosters were to be fitted to give more thrust at height and so Sukhoi decided on a single pure research prototype, designated T-43, to test the new engine, intake and rocket motors together, although when it flew on 10th October 1957 the rockets were still to be fitted. After this new prototype had achieved a speed of 2,200km/h (1,367mph) or Mach 2.06 and, separately, a ceiling of 21,500ft (70,538ft) on jet power alone, all thoughts of booster rockets were dropped.

Eventually more T-43 prototypes were completed, but they differed from the original in having the K-5 AAM combined with a new radar called the TsD-30. The missile, which in service became the RS-2-U, was designed by the Grooshin Bureau. The objective now was to turn the original pure interceptor requirement into an aircraft that, along with a ground-controlled system, formed part of a complete weapon system which became known as the T-3-51. Pre-production T-3/PT-8 aircraft were converted to the new standard and, in due course, the type entered service as the Su-9. For the new fighter the ASCC retained the codename *Fishpot* but called the Su-9 *Fishpot-B* while the earlier T-3 became *Fishpot-A*. The new type suffered quite a number of teething troubles but still served with a large number of PVO (Air Defence Force) regiments before

An early production Sukhoi Su-9.

unacceptable, and so on 1st June 1959 the T-5's test programme was brought to a halt. Alternative versions were considered with side intakes including the T-51 which had semi-circular shaped inlets.

Sukhoi T-47, PT8-4 and Su-11

Further efforts were also made to find a better Almaz radar/shock cone combination and these resulted in yet another prototype called the T-47. This fighter had both the search and tracking antennas housed in a single large intake centrebody and made its maiden flight on 6th January 1958. However, it only completed a few development flights before the AL-7F-1 engine was taken out to go into another prototype, called the PT8-4, which was actually the first production PT-8 converted. This second aircraft also had the Almaz radar, but this time it was combined with an air-to-air missile armament that gave it a higher priority over the T-47, which carried gun armament alone. Further prototypes followed, but these had the Koonyavskiy Oryol (Eagle) radar fitted together with the K-8-2 (K-8M) missile, a new more advanced radar-homing weapon from the Bisnovat Design Bureau which eventually proved to be the USSR's first missile of this type to enter production (in service it was designated R-8). The combined air/ground weapon system's development, undertaken in response to the directive of 16th April 1958 mentioned earlier, proceeded in parallel with the Su-9/T-3-51.

To serve as a prototype the old PT-7 prototype received a new T-47 nose (making it the T47-3) and made its first flight in this aerodynamic form (but without the radar installed) on 25th December 1958. Several more prototypes followed and production aircraft were designated Su-11, the first flying in July 1962 after the whole weapon system had itself been designated Su-11-8M. However, following a fatal crash in October caused by an engine failure, together with some severe criticism directed at the aircraft by Aleksandr Yakovlev, only just over a hundred Su-11s were to be completed. Yakovlev used his influence to promote his twin-jet Yak-28P interceptor (Chapter Four) ahead of Sukhoi's design, citing the increased safety offered in having two power units; he also described Su-11 airframes awaiting delivery at Novosibirsk as 'junk'. Nevertheless those Su-11s that were completed served alongside the Su-9

withdrawals began in the late 1960s. The Su-9 was eventually replaced by the Tupolev Tu-28, Mikoyan MiG-25P and Sukhoi's own Su-15.

Sukhoi T-5

Yet another configuration was studied by this test bed prototype, which introduced two engines instead of one, the engine choice being the Tumanskiy R11F-300. Lyul'ka's AL-7 was reaching the end of the line from a development point of view and the newer AL-9 was experiencing problems, so the Air Force

ordered Sukhoi to try this alternative. The result was a much fatter and more draggy rear fuselage, although the air was still supplied by a T-43-style intake. The prototype itself was actually the old T-3 airframe converted and, as such, it made another maiden flight on 18th July 1958. With the extra power now available to it the T-5 could easily exceed the T-3's top speed, but the engines themselves were really designed to cope with maximum speeds of less than Mach 2. In addition, the aircraft's longitudinal stability was very low and quite

Soviet Secret Projects: Fighters Since 1945

until the early 1980s. In the West it was known as *Fishpot-C*. In service neither the Su-9 nor the Su-11 carried guns.

The long and very complex series of T-43/T-47 designs and prototypes had all evolved from the original T-3, and getting the radar-equipped Su-9 and Su-11 into service required a substantial development effort spread over what was, for the 1950s, a considerable period of time. Their story has been described here because the airframes are tied in so closely with the Su-7 series, but the rest of the Soviet Union's early radar-equipped fighter developments follow in the next chapter. Until the arrival of the MiG-25 in the late 1960s (Chapter Eight) the Su-9 was the fastest of all of the Soviet Union's combat aircraft; it also had the highest service ceiling.

Yakovlev Studies

Yakovlev Yak-135

The existence of this light fighter was accidentally discovered in a photograph. In 1999 a Russian aeronautical magazine published a view showing Yakovlev OKB designers in discussion and behind them, on the wall, was a large poster showing an aircraft entitled 'Yak-135 light tactical fighter'. It had a pointed nose, side intakes, wings with moderate sweep on the leading edge and a straight trailing edge, a similar horizontal tail and a swept vertical tail. The jet nozzle was not clearly visible but left the impression of being placed under a short tail boom. No further information is available and it is not even clear when this project was conceived. It is included here on the supposition that it preceded the Yak-140 but that cannot be confirmed.

Yakovlev Yak-140

Alongside the new supersonic projects produced by Mikoyan and Sukhoi in the mid-1950s came another from Yakovlev, but this OKB's contribution to this part of the story covers just the one design. The Yak-140 was a lightweight single-seat fighter designed by V V Barsukov and was considered by the Design Bureau as a continuation of the series of light fighters described in Chapters Two and Four; in particular it was a direct descendent of the Yak-50.

Go-ahead for the Yak-140 was made through a SovMin resolution issued on 9th September 1953, the preliminary project having been completed the previous July. Two prototypes were authorised and the first had to be ready for its state acceptance trials by March 1955. The covering specification requested 1,650km/h to 1,750km/h (1,025mph to 1,088mph) top speed, service ceiling 18,000m (59,055ft) and a range when flying at 15,000m (49,213ft) altitude of 1,800km (1,119 miles). The design team's calculations suggested that the Yak-140 could meet these limits, and it would carry 1,275kg (2,811lb) of internal fuel. From a relatively small size and low wing loading, the Yak-140's performance estimates suggested a substantial top speed and rate of climb of 12,000m/min (39,370ft/min) at sea level and 1,800m/min (5,906ft/min) at 15,000m (49,213ft) together with good manoeuvrability in both vertical and horizontal planes. It was also expected to operate from unpaved surfaces and had the OKB's favoured bicycle undercarriage.

The intended powerplant was to be a single Mikulin AM-11 engine giving 39.2kN (8,818lb) of thrust dry and 49.0kN (11,025lb) with reheat. Versions were considered with either the Lyul'ka AL-7 or Klimov VK-3 but these gave an altogether heavier aircraft – in fact calculations showed that using the lightweight AM-11 gave an aeroplane of roughly half the weight of the alternative versions yet, with the same armament and equipment aboard, still offering the same flight performance. Manufacturing costs would also be very much reduced while the fuel consumption would be about 50% less. In the event however, the AM-11 was delayed (as noted under the Mikoyan Ye-1 section) and the state trials were postponed until the first quarter of 1956. In due course the first prototype was fitted with an AM-9D, giving 25.5kN (5,730lb) of thrust dry and 31.8kN (7,165lb) with reheat. Also only two Nudel'man 30mm cannon were carried (housed in the lower forward fuselage) when later aircraft were expected to have three guns. As one might have expected the estimated performance figures

with this less powerful engine were reduced, and they included a maximum of 1,275km/h (792mph) at sea level and 1,250km/h (777mph) at 13,000m (42,651ft). The aircraft's dimensions were unaltered but the take-off weight when carrying 1,000kg (2,205lb) of fuel was now 4,500kg (9,920lb).

The Yak-140 prototype was finished late in 1954 and successfully concluded its pre-flight testing, including high-speed taxi runs, to the point where the aeroplane was ready to begin its flight trials. During these runs the machine had behaved as expected and all of its systems had worked well. Flight clearance was given on 10th February 1955 and, because the wind tunnel testing carried by TsAGI had revealed no problems, no speed restrictions were in place. However, by now MAP had decided that fighters of this type should be more powerful and offer more capability; in fact the Ministry had become more interested in the products from Mikoyan and Sukhoi described earlier. As a result the Yak-140 was never allowed to fly and the programme was cancelled by a SovMin resolution dated 28th March 1956 and a MAP order placed nine days later. It appears that the Ministry simply lost interest in the Yak-140, which must have been very frustrating for Aleksandr Yakovlev. His comments have not been recorded, but one suspects that they may have been quite colourful.

The Yak-140's wing was swept back 55° and had a t/c ratio of 6% at the root and 8% at the tips; whether during this period Yakovlev considered a delta wing alternative is, to date, unrecorded. The existence of the Yak-140 appears to have been unknown in the west until about the mid-1990s, well after the end of the Cold War, and the type was to be the last single-seat frontline fighter to come out of the Yakovlev Design Bureau. The more advanced Yakovlev designs from this period were all fitted with a powerful radar and are therefore covered in Chapter Four.

The never-to-be-flown Yakovlev Yak-140 prototype.

Supersonic Fighters – Data / Estimated Data

Project	Span m (ft in)	Length m (ft in)	Gross Wing Area m² (ft²)	All-Up-Weight kg (lb)	Engine kN (lb)	Max Speed / Height km/h (mph) / m (ft)	Armament
Yakovlev Yak-1000 (* = some sources say)	4.52 (14 10) 4.59 (15 0.5)*	11.69 (38 4) without pitot	14.0 (150.5)	2,510 (5,534)	1 x RD-500 15.6 (3,505)	1,100 (684) at height	None
Yak-1000M / Mya M-33	5.6 (18 4.5)	7.5 (24 7)	29.8 (320.0)	3,300 (7,275) 'flying' weight	1 x TRD-5 17.6 (3,970)	1,100 (684) at 10,000 (32,808)	None
Cheranovskii Jet Fighter	8.48 (27 10)	4.67 (15 4)	18.0 (193.5)	1,900 (4,189)	2 jet engines 4.9 (1,100)	800 (497) at S/L	2 cannon
Cheranovskii BICh-26 (approximate data)	8.0 (26 3)	10.1 (33 1.5)	27.0 (290.3)	4,500 (9,921)	1 x AM-5 19.6 (4,409)	1,909 (1,186) Mach 1.7 at 7,000 (22,966)	Unknown
Mikoyan I-350 (flown)	9.73 (31 11)	16.652 (57 7.5)	36.0 (387.5)	8,000 (17,630), Max 8,710 (19,197)	1 x TR-3 44.1 (9,921)	1,240 (771) at S/L, 1,266 (787) at 10,000 (32,808)	1 x 37mm + 2 x 23mm cannon
Mikoyan I-360 1st Prototype (flown)	9.04 (29 8)	13.9 (45 7)	25.15 (270.7)	6,600 (14,550), 7,962 (17,553) w tanks	2 x AM-5A 21.1 (4,740)	1,225 (761) at 3,000 (9,842), 1,202 (747) at 5,000 16,404)	2 x 37mm cannon
Mikoyan MiG-19 Early Production (flown)	9.00 (29 6)	12.54 (41 2) without pitot	25.16 (270.8)	7,500 (16,534), 8,500 (18,739) w tanks	2 x RD-9B 25.5 (5,731), 31.8 (7,164) reheat	1,450 (901) at 10,000 (32,808)	3 x 23mm cannon; 2 rocket pods or 2 bombs up to 250kg (551 lb)
Mikoyan I-370 (flown)	9.0 (29 6)	? As first proposed 10.44 (34 3)	25.16 (270.5)	7,030 (15,498)	1 x VK-7F 41.1 (9,260), 61.4 (13,825) reheat	1,483 (921) at 11,000 (36,089)	1 x 37mm + 2 x 23mm cannon; 2 rocket pods or 2 bombs up to 250kg (551 lb) bombs
Mikoyan SM-12 (flown)	9.00 (29 6)	13.66 (44 10) cone forward but without pitot	25.16 (270.8)	7,654 (16,873), 8,696 (19,171) w tanks	2 x R3-26 37.2 (8,375) reheat	1,840 (1,143) at 11,000 (36,089), 1,926 (1,197) at 12,500 (41,010)	2 x 30mm cannon; 2 rocket pods or 2 bombs up to 250kg (551 lb)
Mikoyan Ye-2 (flown)	8.11 (26 7)	13.23 (43 5)	21.0 (226.0)	5,334 (11,759)	1 x AM-9B 25.5 (5,730), 31.8 (7,165) reheat	1,920 (1,193) at 11,000 (36,089)	2 x 30mm cannon
Mikoyan Ye-2A (flown)	8.11 (26 7)	13.23 (43 5)	21.0 (226.0)	6,250 (13,779)	1 x RD-11 37.2 (8,375), 50.0 (11,245) reheat	1,900 (1,181) at height	2 x 30mm cannon
Mikoyan Ye-4 (flown)	7.75 (25 5)	13.23 (43 5)	23.15 (249.2)	5,200 (11,464)	1 x AM-9B 25.5 (5,730), 31.8 (7,165) reheat	1,296 (805) at 10,500 (34,449)	2 x 30mm cannon
Mikoyan Ye-5 (flown)	7.75 (25 5)	13.23 (43 5)	23.15 (249.2)	6,250 (13,779)	1 x RD-11 37.2 (8,375), 50.0 (11,245) reheat	1,970 (1,224) at 10,500 (34,449)	2 x 30mm cannon
Mikoyan Ye-6/MiG-21 (flown)	7.154 (23 5.5)	13.46 (44 2)	23.0 (247.6)	6,915 (15,245)	1 x R-11F-300 38 2 (8,600), 56.2 (12,655) reheat	2,174 (1,351) at 12,500 (41,010)	1 x 30mm cannon, R-3S (K-13) AAM
Sukhoi Su-7 (flown)	9.31 (30 6.5)	16.51 (54 2)	34.0 (365.6)	10,859 (23,940) with tanks	1 x AL-7F 67.1 (15,100), 87.7 (19,730) reheat	1,250 (777) at S/L, c2,264 (1,407) at 11,000 (36,089)	3 x 30mm cannon, 2 x 8 pods unguided rockets
Sukhoi T-3 (flown)	8.54 (28 0)	17.07 (56 0)	34.0 (365.6)	11,110 (24,493)	1 x AL-7F 73.5 (16,535), 8.0 (22,045) reheat	2,100 (1,305) at 12,000 (39,370)	2 x 30mm cannon (never fitted), 2 x K-7L or K-6V AAM
Sukhoi Su-9 (flown)	8.54 (28 0)	18.055 (59 3) with pitot	34.0 (365.6)	12,512 (27,584) with tanks	1 x AL-7F-1 66.6 (14,990), 94.1 (21,165) reheat	2,230 (1,386) at 12,000 (39,370)	4 x RS-2-US AAM
Sukhoi T-5 (flown)	8.54 (28 0)	18.38 (60 3.5)	34.0 (365.6)	11,300 (24,912)	2 x R-11F-300 41.2 (9,260), 60.0 (13,490) reheat	2,120 (1,317) at 13,000 (42,651)	None carried
Sukhoi Su-11 (flown)	8.54 (28 0)	17.55 (57 7) with pitot	34.0 (365.6)	13,986 (30,833) with tanks	1 x AL-7F-2 66.6 (14,990), 99.0 (22,265) reheat	2,340 (1,454) at 12,000 (39,370)	2 x R-8M AAM
Yakovlev Yak-140	7.395 (24 3)	13.34 (43 9)	20.0 (215.1)	4,850 (10,692)	1 x AM-11 39.2 (8,818), 49.0 (11,025) reheat	c1,700 (1,057) at height	3 x 30mm cannon. Capability to carry rocket or 200kg (441 lb) bombs

Chapter Four

Fighters with Radar

The 1950s saw probably the greatest advances in fighter design since the dawn of the jet age. Besides the improvements in aerodynamics and engines, which took the speed range from about Mach 1 at the start of the decade to at least Mach 2 by the end, there were concurrent developments in the equipment and weapons carried by these aircraft. The air-to-air missile (AAM) was a big step forward, although the early generations of missiles were quite limited in what they were capable of achieving. Missile designers worldwide had to overcome many development problems, but a weakness of some early Soviet AAMs was that the fighter had to stay on the level to hit its target. If the target banked the missile would fall out of the radar cone beamed onto it by the defending fighter.

Recognition has already been given to the Soviet Union's need to produce fighters, or rather interceptors, that could deal with reconnaissance aircraft like the U-2 flying at very high altitudes. However, not only was it

hard work getting some of these early supersonic fighters up to heights well in excess of 15,240m (50,000ft) to make their interceptions and attacks, but there were also problems operating at such heights, particularly concerning stability and manoeuvrability. During the late 1950s and early 1960s manoeuvring to make an attack on a high-flying bomber could be extremely difficult because certain bombers, and in particular the British Avro Vulcan, could out-turn any fighter at heights around 50,000ft and more; this presented real problems to the Soviet fighter designer. Yet, within a few years the advent of surface-to-air missile systems (SAMs) had forced enemy bombers to make at least part of their attacks at low level, because suddenly they were very vulnerable when flying at height. In consequence fighters now had to be developed that were capable of making interceptions at low level.

The key elements for an interceptor fighter were its missiles and radar. The two pieces of

The Sukhoi T-49 prototype with its most unusual air intake arrangement.

equipment had to work together and the weakest element of the system would set the capability of the aircraft as a whole. However, a good radar was critical for finding a target. Coverage has already been given to the Sukhoi Su-9 and Su-11, which introduced the concept of a weapon system that combined both air and ground-based radars to help the fighter find its target. One of the earliest Soviet radars was the Toriy (Thorium) set developed by a team based at the NII-17 Research Institute in Moscow and led by A B Slepooshkin, the pioneer of Soviet radar technology. The development of early warning detection and tracking radars, like the Toriy, began in earnest in 1946, but by the mid-1950s more sophisticated equipment was becoming available which needed the right aircraft to carry it. This chapter takes a look at the first

The first Lavochkin 'Aircraft 200' prototype was to have had its radar housed in a centre-intake radome.

Cutaway drawing of the original Lavochkin '200's internal arrangement. Note the split intake ducting to the two engines.

radar-equipped fighter designs of Lavochkin, Mikoyan, Sukhoi and Yakovlev produced the late 1940s, and also the fighters that followed the Sukhoi Su-9/Su-11 series in the mid 1950s and the 1960s.

From 1941 the defence of the Soviet Union's own territory was put into the hands of an organisation known as *PVO Strany* (the National Defence of the Homeland), which in 1982 was renamed *Voyska PVO*. It was usually commanded by an artilleryman and combined an assortment of surface-to-air defences (missiles and anti-aircraft artillery or AAA), early-warning and ground-based radars and interceptor fighters. After the end of the Great Patriotic War, the Kremlin realised that the country would quickly become very vulnerable to attack by bombers carrying nuclear weapons and so a much more comprehensive combination of defensive facilities had to be created and put in place. This included the fighters and their equipment described here, but also new ground-based systems plus an efficient communications system to tie it all together. The pace of development for new facilities aimed at the destruction of enemy jet bombers was hectic, and remained so throughout the Cold War – most of the fighters described here were just part of that massive effort. Note that the offensive part of Soviet aviation was controlled by the VVS, the Air Force of the USSR, which operated fighter-bombers such as the Sukhoi Su-7B as well as its own fleet of heavy bombers.

The First Radar Fighters

In January 1948 an official requirement was issued for an advanced radar-equipped fighter. It was actually described as a cover interceptor and was intended to serve as a night and all-weather fighter to oppose any type of enemy aircraft breaking into Soviet airspace. A crew of two was requested and the document brought responses and experimental prototypes from all four of the main fighter Design Bureaux.

Lavochkin 'Aircraft 200' and '200B'
Lavochkin's early studies embraced several powerplants, two Derwent Vs (which were

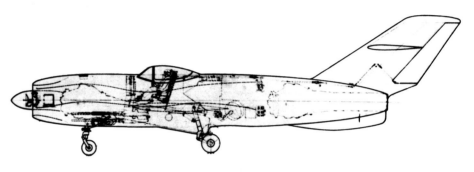

not powerful enough), two RD-45F engines or a single TR-3; final choice was a pair of VK-1 units. Aerodynamically this project was not particularly advanced having a thick wing swept back at 40° and a t/c ratio of 9.5%. It did introduce, however, an unusual engine arrangement with the relatively fat centrifugal VK-1s placed in tandem and fed by a circular nose inlet. The air ducting was split into three with the two outer channels passing either side of the cockpit to the rear engine and the centre channel dropping down to feed the foremost unit. The crew sat side-by-side and 'Aircraft 200' used a tricycle undercarriage with a quite narrow wheel track and twin wheels on each main gear. Armament comprised three Nudel'man N-37 37mm cannon in the lower nose, one to port and two to starboard, and a Toriy radar was to be housed in a radome mounted in the centre of the intake. The mock-up was approved on 24th February 1949 and the first prototype, labelled '200-01', made its first flight on the 9th September. There was no radar installed but the aircraft achieved Mach 0.946 on the level and Mach 1.01 in a dive.

The second machine, '200-02', did get its radar, a Toriy-A set, but in a new housing fitted to the upper part of the inlet. There were

other changes, including a small 'keel' under-fin, and the aircraft showed a slight improvement in performance. Official flight testing was completed in April 1951 and the '200' was in fact the only one of this series of interceptors to pass all of its state trials, so recommendations were made that the type should be put into production. A go-ahead was, however, never given and seven months later a replacement specification was issued which pushed up the range and endurance by a substantial margin (to allow standing patrols to be made). This also requested a heavier radar with a longer detection range and, as a result, work on 'Aircraft 200' in its original form came to a halt.

It was replaced by a third aircraft, designated 'Aircraft 200B', which introduced some changes to meet these additional requirements. The radar, a new and more advanced type called the RP-6 Sokol (Falcon), needed a much larger radome which in turn required alterations to the intake arrangement (trials had shown that overall the Toriy-A was unreliable and underdeveloped). Uprated engines were also fitted and to feed them there were now three inlets, the bottom supplying the lower front engine while two more, of an 'elephant ear' type, were set on the sides of the

The second 'Aircraft 200' prototype did receive its radar, but in a new radome mounted on the upper intake lip.

An increase in the flight endurance requirements and a different (Sokol) radar meant that the third Lavochkin 'Aircraft 200', designated '200B', had an altogether different nose.

Cutaway drawing of the '200B's internal arrangement.

This view of the Lavochkin '200B' prototype shows the complex air intake surrounding the large radome.

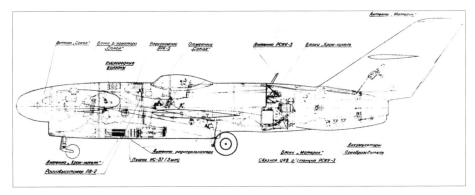

upper nose to pass through to the rear power unit. The nosewheel was moved forward, each main gear had just one large wheel and there were now two ventral fins. A new wing was also fitted that looked very similar to the old form but which used a different method of construction and housed more fuel.

The '200B' flew on 3rd July 1952 with a mock-up radar installed. The real equipment was used for the first time during a flight made on 10th September. The test programme embraced 109 flights but, all in all, the 'Aircraft 200' series, also known as the La-200, presented another disappointment to Lavochkin in that it failed to gain a production order. By now a new prototype from Yakovlev was also under test, the Yak-120 (below), and this was accepted for squadron service. In November 1953 orders were given to end testing on the '200B' and fit the Sokol radar into the Yak-120. The two versions of 'Aircraft 200' took 2.6 and 2.8 minutes respectively to reach 5,000m (16,404ft) and their ceilings were 15,150m (49,705ft) and 14,125m (46,342ft).

Mikoyan I-320

In Chapter Two a brief mention was made of the Mikoyan SP-1 and SP-2 variants of the MiG-15 and MiG-17 respectively, which were among the first attempts by this OKB to produce fighters equipped with a powerful intercept radar. These prototypes were flown in April 1949 and March 1951 and the radars that they used were the Toriy (Thorium) and the Korshoon (Kite) respectively; the latter, like the Toriy, was also developed by the Slepooshkin design team. However, the huge workload that these radars created for the pilot made interceptions far too difficult – both fighters were single-seaters and Toriy did not have an automatic target-tracking facility. Consequently, the SP-1 and SP-2 were classed as failures and neither entered production, although much had been learnt about radar operations in such aircraft.

Concurrent with these projects was Mikoyan's response to the January 1948

requirement, the I-320 or 'Izdeliye R'. Like Lavochkin the Mikoyan designers accepted that the job of operating the radar would require a second crewman, who in the I-320 was seated alongside the pilot. He would add weight of course, but would deal with the navigation and get the fighter close to its target, so the pilot only had to worry about flying and fighting. The '200's twin-engine tandem arrangement was also used with air supplied by a nose intake. The forward engine was placed underneath the cockpit (the underbelly jetpipe was level with the rear of the cockpit) while the second unit was housed in the rear fuselage and had a tail jetpipe. Again the ducting was split into three with the middle section feeding the front engine and the outside pair going to the rear unit. Although the I-320 was designed from scratch, the aerodynamics had much in common with the MiG-15, with the same wing fittings and 35° of sweepback, and the airframe itself employed established manufacturing practice. A Toriy-A radar was to be housed in a radome placed in the upper nose above the intake and two 37mm cannon were carried, one on each side of the forward fuselage just below wing level.

Two prototypes were built, designated R-1 and R-2, and the first of these flew on 16th April 1949 without its radar in place. During this period Mikoyan built a reputation for having the ability to produce a new prototype ready for flight trials in the shortest possible time, which accounted in part for the R-1's first flight taking place nearly five months earlier than the Lavochkin '200' (although the

Sukhoi Su-15 below beat them both). R-1's flight characteristics proved to be reasonably good although some early problems experienced in the MiG-15 were repeated by this design and proved difficult to solve. These were mainly lateral instability at Mach numbers between 0.89 and 0.90 and a tendency to fall into a bank at instrument speeds between 840km/h and 930km/h (522mph and 578mph).

R-2 flew in December and had the Toriy-A fitted plus more powerful VK-1 power units offering 26.4kN (5,950lb) of thrust each. After a gun-firing accident had damaged the R-2's nose the Toriy was replaced by the more advanced Korshoon and other modifications were made to the airframe to improve the transverse instability, including reducing wing anhedral from 3° to 1.5°. As such the aircraft was redesignated R-3 and flew again on 31st March 1950, but by the middle of that year it was clear that the type would not enter production (in part because of the problems experienced with the radar). With the new powerplant the R's maximum take-off weight was 12,095kg (26,664lb) and top speed 1,090km/h (677mph) at 1,000m (3,281ft) and 1,060km/h (659mph) at 10,000m (32,808ft). The respective service ceilings for the two versions were 15,000m (49,213ft) and 15,500m (50,853ft) and R-1's time to 5,000m (16,404ft) was 2.3 minutes.

Sukhoi Su-15
The Soviet Union's Council of Ministers experimental aircraft manufacturing plans established in March 1947 included a new

The first prototype Mikoyan I-320, also known as the R-1.

interceptor from the Sukhoi Design Bureau, which was initially designated 'Aircraft P' and later Su-15; this was actually Sukhoi's answer to the radar-equipped fighter requirement. Once more an unusual design resulted in that the two RD-45 power units were again mounted in tandem, one low in the middle fuselage and the second in the rear fuselage (originally it had been intended to use Derwent V units). Each engine had its own jetpipe, the first exhausting beneath the middle fuselage, and the cockpit was offset to port so that the rear engine duct could be offset to the starboard side of the intake. This was possible because, unlike the Lavochkin and Mikoyan offerings, Sukhoi's fighter was a single-seater. A large upper nose radome housed a Toriy radar.

A 30° swept wing was used, in part based on information supplied in May by TsAGI, and the preliminary project and a full-scale mock-up were completed at the end of the year. These were finally approved in the summer of 1948 and the prototype was rolled out in October. The Su-15 finally flew on 11th January 1949, after suffering the ignominy of having the nosewheel leg collapse when attempting to make its first flight on 10th November 1948. Then, during a flight on 3rd June 1949, the fighter began to experience flutter and had to be abandoned by its pilot; as a result the second machine was never completed and the programme was cancelled. An Su-15 trainer project was also proposed with a longer

The Su-15 could not be considered an attractive aircraft but it did present an impressive appearance.

The first Yakovlev Yak-50 prototype.

cockpit to accommodate a second seat for an instructor and a second Toriy radar display. It was to be armed with one 37mm cannon and a 12.7mm machine gun but was never built. The original prototype exhibited a service ceiling of 15,000m (49,213ft) and took 2.5 minutes to reach 5,000m (16,404ft).

Summarising the three designs, the Lavochkin '200' series and the Mikoyan I-320 heavy fighters were far from glamorous aircraft, but at the time of their appearance the unique tandem engine arrangement and side-by-side two-seat cockpits were new ideas. Likewise the first aircraft to receive the Su-15 designation was quite similar but has always been associated with a crash that contributed to the closure of the Sukhoi OKB. As a result none of them have received the level of attention that they deserve, yet they form an important part of the Soviet fighter story. It is fascinating that all three OKBs selected the same engine layout, yet this arrangement was never again used by any Soviet or Russian fighter design that was to be turned into hardware. None of these aircraft were to receive their official certification (although the I-320 and '200' underwent state acceptance testing in 1950 and 1951) and any production plans

were quashed by updates and improvements made to the requirements. However, the real killer was probably the appearance of the new Yak-120 design from Yakovlev described shortly. This proved to be the successful project in the quest to give the Soviet Union a radar-equipped fighter.

Yakovlev Yak-50

The final design tendered to the original 1948 competition differed from the others in that it was a much smaller and lighter aircraft and had just one engine, a 26.4kN (5,950lb) VK-1; it was also a single-seater. In many respects

this was as a stretched derivative of the Yak-30 described in Chapter Two but its wings were swept 45° at 1/4-chord, the t/c ratio was 12% and two 23mm cannon were fitted in the lower nose beneath the intake. The radar was a Korshoon with a single parabolic scanner housed on the upper intake lip. Unlike the Yak-30, which had a tricycle undercarriage, the Yak-50 used a zero-track bicycle arrangement with twin main wheels plus outriggers right on the wingtips. This format had earlier been tested as a non-retracting installation by one of the Yak-25 straight-wing prototypes described in Chapter One.

The first Yak-50 prototype was first flown on 15th July 1949 by Sergei Anokhin, who had already flown the I-320 and Su-15 (the Lavochkin '200' was still to fly). Performance and handling proved to be excellent, despite having just the single power unit, but the fighter's weaknesses included not having 37mm guns in its armoury. Again the problem of the pilot on his own operating this rather complex and inadequate early-generation radar reared its head – the job was just too much to take on and another crew member was definitely required. Two more prototypes followed but the Yak-50 proved difficult to land in a moderate crosswind and very difficult to control on a wet surface. In addition, above Mach 0.92 it suffered from lateral oscillation, which did not help with gun aiming, so the programme was abandoned on 30th May 1950. However, this proved to be Yakovlev's first supersonic fighter and it achieved Mach 1.03 or 1,065km/h (662mph) on the level at 10,000m (32,808ft). The Yak-50 also needed just 1.5 minutes to reach 5,000m (16,404ft) and had a service ceiling of 16,600m (54,462ft).

–

The next part of the radar fighter story describes Yakovlev's success in producing the large twin-engined Yak-120 interceptor prototype, which eventually entered service as the Yak-25. However, the 120 'evolved' from, and was linked to, the Yak-50 through a couple of unbuilt designs.

Yakovlev Yak-60

This design was actually a relatively light single-seat fighter project fitted with a 45° swept wing. It shared the Yak-50's overall configuration and was powered by a VK-1 engine, but it was a larger aircraft with a longer fuselage and greater wing area; the fuselage itself had been stretched from 9.465m (31ft) in length to 11.0m (36ft 1in). A tricycle undercarriage was fitted and, besides the internal fuel, the aircraft could carry large external tanks under its wings. Aleksandr Yakovlev

approved the Yak-60 design and layout on 20th November 1948 and, from the following December into January 1949, scale models of the design were assessed by TsAGI in its T-106 wind tunnel. These were tested at speeds up to Mach 0.9 and the results indicated that the Yak-60, had it been built, could have exceeded this Mach number. A second version of the Yak-60 was proposed with a bicycle undercarriage and a radar in a fairing mounted above the air intake.

In due course, however, the project was abandoned but in 1952 the aerodynamic research data acquired from it, and the fighter's basic layout, were worked into the new Yak-120. The latter was, as noted, a twin-engined aeroplane, but the Yak-60's external tanks were 'replaced' by two AM-5 engines, which left additional space in the fuselage because the VK-1 could now be taken out. As such, the Yak-60 formed an important link between the Yak-50 and the Yak-120.

Yakovlev Yak-13

An immediate predecessor of Yak-120, and a twin-engined design with jet engines mounted under the wings, work started on this interceptor project in August 1951. Hardly any information is available but the Yak-13 had a wingspan of 10.6m (34ft 9in), length 15.4m (50ft 6in) and wing area 28.0m^2 (301.1ft^2), figures that approach those of the eventual Yak-120 prototype. This was the second use of the designation by the Yakovlev OKB, the first having been applied to a light touring aircraft of 1946.

Yakovlev Yak-120/Yak-25

In 1951, after the failure of their more recent single-engined fighter prototypes, Aleksandr Yakovlev and his team concentrated on a new twin-engined two-seat patrol interceptor design as a private venture (in other words at the OKB's own risk). However, official support soon arrived because on 6th August, at a meeting chaired by Stalin himself, orders were placed for prototypes of the Yak-120 (the same meeting also cleared the way for the Mikoyan I-360 which led to the MiG-19). One aspect of the Soviet Union's air defence programme that was giving planners a headache was finding a way of covering the Far East sector of the country, a part of the world which suffered from extremely cold temperatures and severe weather condi-

tions. Organising a system of surface-to-air missiles to embrace the whole of this area was a big problem, so an interceptor having the capability to make long patrols and fitted with a good navigation system was seen as an ideal solution.

The engine choice was the Mikulin AM-5 which offered 19.6kN (4,410 lb) of thrust (it did not have reheat and was later renamed the RD-5A), and the Yak-120 was eventually selected to carry the RP-6 Sokol (Falcon) radar. The Sokol, which was designed by the G M Koonyavskiy Design Bureau, was showing more promise than previous equipment and offered a detection range of 30km (18½ miles). In addition placing the engines in underwing nacelles also made room for a massive radome on the nose.

The Yak-120's layout differed markedly from all of the aircraft industry's previous attempts to create a successful radar-equipped fighter (in timescale this design came much earlier than the Su-9 in Chapter Three). Its wings were swept 45° and had parallel leading and trailing edges, and the nacelles were slim and had no room available to house an undercarriage as well as the engine. There was also no room in the wings and so a bicycle arrangement with outriggers was employed, but this held no worries for Yakovlev because the Bureau had already used this type on some of its previous designs. Two N-37L cannon were placed low on the sides of the centre fuselage and, overall, the aircraft had a relatively light structure for its size and the equipment it was intending to carry; in fact it was smaller and lighter than the Lavochkin '200B'.

Having suffered the problems with the earlier attempts at finding radar-equipped fighters, the PVO was not convinced that this new type could be in service without any prolonged delays. Nevertheless, despite these misgivings, a directive approving a go-ahead to the Yak-120 programme was placed on 10th August 1951; this document also requested work to proceed on the Yak-125, a photo-reconnaissance aircraft which used the same airframe. Two Yak-120 prototypes were built and the first of these flew on 19th June 1952. Early trials showed that nearly all of the specified requirements had been met and the aircraft could reach 10,000m (32,808ft) in 4.3 minutes, although it was a little short on range. Service ceiling was 14,500m (47,572ft).

Yakovlev's new design carried the day and put to waste the earlier efforts made by Lavochkin, Mikoyan and Sukhoi. In due course it

entered service as the Yak-25, but delays in getting the radar to work reliably meant that the first examples had the RP-1 Izumrood set as an interim fitting. It was a test programme carried out in 1953 using the La-200B prototype which really helped to get the Sokol to work properly and the final radar trials using the upgraded Yak-120 were completed in April 1954. The Sokol variant officially entered service as the Yak-25M, but to most it was still called the Yak-25. The first units to get the type were PVO regiments based beyond the Arctic Circle in the most northerly regions of the USSR; the West allocated the codename *Flashlight* and the fighter served until the end of the 1960s.

Other versions to be built included the Yak-125 tactical photo-reconnaissance prototype, the Yak-125B tactical nuclear strike aircraft prototype, the Yak-25R reconnaissance aircraft, the Yak-25RV high-altitude reconnaissance aircraft (with very long span wings) and the Yak-25K interceptor. There was also an AM-11-powered proposal (below), and later a family of supersonic types based on a new design (the Yak-121), which incorporated experience gained with the Yak-25. These are described later.

Yakovlev Yak-25K

This variant introduced missiles, namely the Grooshin K-5, as part of the USSR's programme to put AAMs into frontline service on several different fighter types. The Sokol radar was replaced by the RP-1-U Izumrood and four missiles were carried, two under each wing inboard of the engine nacelles; the full weapon system was called the Yak-25K-5. The first Yak-25K was a converted Yak-25M and flew in 1955, but only a few were built and most of these served as test aircraft for other missile systems, including the Yak-25K-75 prototype that tested the Toropov K-75 AAM. In fact during this period the work

undertaken in the Soviet Union to get fighters into service armed with air-to-air missiles was colossal. It followed a blanket directive issued on 30th December 1954 ordering the various fighter OKBs to adapt and test their current aircraft types as interceptors. One design produced in response to this was the MiG-17PFU missile-armed version of Mikoyan's successful fighter.

Yakovlev Yak-2AM-11

Another version of the Yak-120/Yak-25 was this two-seat supersonic long-range interceptor proposal, which was also to be developed into a tactical reconnaissance derivative. The key was the introduction of a new engine, the Mikulin AM-11 offering 39.2kN (8,820 lb) thrust dry and 49.0kN (11,025 lb) with reheat. A SovMin resolution was passed on 10th June 1954 which demanded that the interceptor should have a top speed of at least 1,350km/h to 1,400km/h (839mph to 870mph) at 10,000m (32,808ft) altitude (that is, well above Mach 1), a top speed on dry thrust of 1,200km/h (746mph), a service ceiling in reheat of between 17,000m and 18,000m (55,774ft and 59,055ft), a time to 10,000m (in reheat) of 1.8 minutes and maximum range with external fuel tanks 3,000km (1,865 miles). The reconnaissance version's top speed (in reheat) and range was to be higher, approaching 1,500km/h (932mph) and 4,000km (2,486 miles) respectively, but the specified ceiling and climb rates were roughly the same. It would carry a battery of four cameras plus one 30mm cannon while the interceptor was to be fitted with an RP-6 Sokol radar, three 30mm guns and (in the overload condition) a selection of rocket projectiles.

State acceptance trials on the interceptor prototype were due to start during the last quarter of 1955 with those for the reconnaissance version following about six months later. Unfortunately, the AM-11's develop-

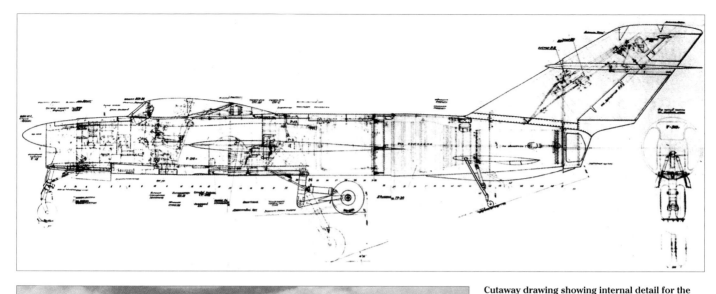

Cutaway drawing showing internal detail for the Lavochkin 'Aircraft 190'.

The sole 'Aircraft 190' prototype flew on just eight occasions.

ment was delayed and priority was also given for the first examples to go in the Mikoyan MiG-21F (Chapter Three). Consequently, no engines would be available in time for the Yakovlev design and so on 30th March 1955 another Council of Ministers directive cancelled the Yak-2AM-11. The project's short life was terminated before the allocation of a new Yak-series designation, but the directive also stated that a new development should be started fitted with two reheated RD-9 (AM-9) engines, which became the Yak-121.

–

The Soviet aircraft industry now moved on to more advanced airframes and more capable designs that also introduced supersonic performance. The Sukhoi Su-9 and Su-11 from the previous chapter can be grouped with these machines, together with several of the heavy interceptors to be discussed in due course in Chapter Five.

Lavochkin Studies

Lavochkin 'Aircraft 190'

This aircraft resulted from a direct approach to Stalin by Semyon Lavochkin, who proposed an all-weather interceptor that could attack enemy bombers flying at heights between 15,000m and 16,000m (49,213ft and 52,493ft). The fighter would have excellent manoeuvrability, particularly in the vertical plane, to allow it to get above the bombers and initial estimates suggested that it would be capable of reaching 10,000m (32,808ft) in about one and a half to two minutes and have a ceiling in the range of 19,000m to 20,000m (62,336ft to 65,617ft). Curiously, it was intended to drop small bombs, possibly on parachutes, into the path of the oncoming bombers and thus deny them the airspace to proceed to their target. As work continued the project and its performance were refined with time to 10,000m on dry thrust increasing to three minutes and a

ceiling of 16,500m (54,134ft); in reheat the figures were about 2.2 minutes and 18,000m (59,055ft). The powerplant was a single Lyul'ka TR-3A (AL-5), estimated top speed at 10,000m 1,225km/h (761mph) and range on internal fuel 1,000km (622 miles). A version powered by the Klimov VK-1 was also assessed which offered a lower performance but greater engine reliability.

The result was 'Aircraft 190' which during 1949 was turned from a draft proposal into a preliminary project. It was intended to have a Korshoon AI radar housed in a radome mounted in the upper section of its nose intake, but in fact the radar itself was never fitted. The chosen layout represented a completely new arrangement for the Lavochkin OKB with the wing swept 55° at the leading edge and having a t/c ratio of 6.1%. A quite unusual tailplane of near delta shape was placed towards the top of the fin (with the rudder above and below), two 37mm cannon were mounted under the nose (an alternative three 23mm arrangement was not adopted) and a bicycle undercarriage was used with light outrigger stabilising legs and wheels. The internal fuel load of 'Aircraft 190' totalled 2,100 litres (462gal), which was housed in both the wings and the fuselage.

Design work progressed throughout 1949 and the prototype was rolled out on 3rd February 1951. The first flight was made a week later but problems with the engine quickly appeared. For example, during a taxi run in early March, the AL-5 did not respond soon enough to a reduction in throttle and the air-

Top: **Model of the Lavochkin '190'.** John Hall

Centre and bottom: **This model represents an early version of the Mikoyan Ye-8 before the design was given canards. It appears to be carrying two versions of the K-13 missile.**

frame received some damage. Repairs lasted until the end of May but then an in-flight engine failure on 16th June resulted in a forced landing. Flying resumed in mid-July but only eight sorties were completed before the programme was abandoned on 20th August. Flight test data had revealed that the stability and controllability of 'Aircraft 190' were satisfactory up to speeds of 820km/h (510mph) at 7,000m (22,966ft) altitude and its control characteristics were fine; in addition the aircraft's systems and mechanics (including the hydraulic booster intended to prevent aileron reversal) all worked well. The reliability of the engine was the only problem, but it was serious enough to bring closure to the programme and neither the aircraft nor its engine were to enter production.

Sadly, because of this the '190' has become a somewhat neglected and forgotten design, yet in appearance it shared some of the advanced features used by the concurrent Mikoyan and Sukhoi designs. One reason for this is that the Lavochkin OKB ceased building aircraft a long time ago, many years before the end of the Cold War. Its performance figures included a speed of 1,190km/h (740mph) or Mach 1.03 at 5,000m (16,404ft) (achieved on one test flight), a time to 5,000m of 1.5 minutes and a service ceiling of 15,600m (51,181m).

Mikoyan Studies

Mikoyan Ye-8
One of the most interesting prototypes produced during this period was this much altered and highly advanced development of the MiG-21 (Chapter Three). By the time the acceptance report clearing the way for the MiG-21F to enter service was signed in 1959, the Mikoyan Design Bureau had moved on to the next generation of fighters. Studies were made in two different directions, a heavy interceptor family beginning with the Ye-150 (described in Chapter Five) and the radar-equipped development of the MiG-21 covered here. The 'heavy family' were to be capable of speeds well in excess of Mach 2.5 and would have a nose intake arrangement, which at the time was considered to be the optimal for such speeds.

The development of more advanced radars was progressing well but late-1950s technol-

ogy, with the arrival of microprocessors and other advanced electronics still years away, meant that the resulting equipment could be rather large. As a result there was insufficient room to put a big powerful radar inside the centre shock cone of a Ye-150-type aeroplane. In addition, such high speeds generated a considerable amount of kinetic heat that produced undesirable effects on the inlet ducting, so the solution was to move the intakes somewhere else and fit a large enclosed radome on the nose. Eventually the lateral or 'chin' position was found to be the most favoured choice for the intake and this format, coupled with a nose-mounted radar, became the second new line of development. The basic MiG-21 fighter possessed enormous development potential and the Ye-8 was to be just one of many versions and ideas produced from it over many years.

By 1960 the full project development of a new fighter could begin and the resulting Ye-8 was to make maximum use of MiG-21 experience. In other words it would retain the MiG-21's high level of manoeuvrability, small size and weight but now perform as a complete weapon system capable of dealing with targets in both forward and rear hemispheres in all weathers, day and night. In addition such attacks were to be possible at all heights up to 18,000m (59,055ft) and at speeds ranging between Mach 0.5 and 1.8. Finally, because it was small (little bigger than a MiG-21), it would be capable of mass production and thus replace with relative ease the MiG-21PF on the production lines. The main elements of the weapon system itself, which was known as the S-23, comprised the new and more powerful impulse-emission Sapfir-21 (Sapphire) radar from the Volkov Bureau (which in service was called RP-22) and the Bisnovat K-23 AAM. Work on the latter, a long-range weapon with alternative infra-red homing or semi-active radar guid-

ance systems, had only just begun. The Ye-8 became another type to receive the provisional service designation MiG-23.

The Ye-8's intake received considerable attention before it was selected as the best choice for limiting pressure losses during manoeuvres. In contrast, because of the data available from the MiG-21, the aerodynamics needed less work, but the Ye-8 did introduce another new and quite innovative feature in the form of a 2.6m (8ft 6¼in)-span canard foreplane. This was not a control surface but rather was intended to improve the fighter's longitudinal control. At speeds below Mach 1 it would be able to pivot freely on its axis, but at supersonic speeds would be locked mechanically at a neutral angle of incidence. The mechanism required to freeze the canard proved successful and the new surface did not affect the aircraft's handling characteristics; in fact at speeds between Mach 1.5 and 2.0 it would double the lift co-efficient.

Another change was the Metskhvarishvili R21F-300 engine, developed by the Tumanskiy OKB out of the MiG-21's R11-300, which offered 46.0kN (10,360lb) of thrust dry and 70.6kN (15,875lb) with reheat. In addition, the Ye-8's structure differed quite a bit from the MiG-21 and it could carry more weapons and fuel. As the design work proceeded, the differences between the Ye-8 and its predecessor became more substantial until the only parts common to both the MiG-21PF and the new type were the wings and tail, main gears and some equipment. Ye-8 also introduced integral fuel tanks, which reduced weight but increased fuel capacity, and it became the first Mikoyan product to have them.

Ye-8 was launched by an official decree issued in 1961 and the final drawings were approved by Mikoyan in June of that year; however, delays with the Sapfir-21 meant that an alternative TsD-30TP (RP-2) set had to be

fitted in the prototype. Taxi trials began in late March 1962 and the first aircraft made its first flight on 17th April. A lot of testing was required to make the intake system work, regular problems being experienced with compressor stalling. The aircraft was lost on 11th September during an air test when a compressor disc broke up at Mach 1.7.

By then the second Ye-8 was in the air, having flown on 29th June without radar but with pylons for two short-range K-13 AAMs, but the crash brought all flight-testing to a halt. It was never restarted, the programme being cancelled by a MAP order. The key weakness was the engine, which had only six compressor stages and never reached maturity, and it too was abandoned, to be replaced by the 99.9kN (22,485lb) R27-300 and 122.5kN (27,555lb) R29-300 engines. These introduced more stages with a lower compression ratio on each, and twin spools, and were fitted in the aircraft that finally became the MiG-23 (Chapter Eight); that aircraft also gave a home to the K-23/R-23 missile.

In 1962, before the Ye-8 programme was finally terminated, the Mikoyan design team got to work on a modified Ye-8M proposal complete with the S-23P weapon system (which also still used the MiG-23 label). The low-set intake and engine were modified but the wing, tail and landing gear were retained, although the wing was larger. This project was not built and another rejected proposal was a STOL design of 1963 called the Ye-8F that had additional RD-36-35 lift engines. The Ye-8 itself had a Mach limit of 2.1, a service ceiling of 20,300m (66,709ft) and could reach 18,000m (59,055ft) in 5.9 minutes. Its loss was much regretted by most of Mikoyan's design staff, who regarded the type as their design for the future. It was probably a touch ahead of its time but was an impressive aircraft that might have become an agile air combat fighter, rather like today's MiG-29 or American F-16 Fighting Falcon. The existence of this aircraft was unknown in the West until long after cancellation, so no ASCC codename was ever allocated.

Sukhoi Programmes

Sukhoi T-39 and T-49

One of the studies made to try and improve the ceiling of the T-3 prototype in the previous chapter had been the introduction of water

injection into the engine afterburner. To eval-
uate the idea a prototype called the T-39 had
been started, which in fact was the third pre-
production T-3 modified. However, construction work was stopped before June 1958 and
before the aircraft was ready to fly. Problems
had also arisen with the aerodynamics of the
T-47 prototype's nose arrangement (also
described in Chapter Three), due in part to the
large radome. Therefore yet another prototype, called the T-49, was produced against a
GKAT order of 6th August 1957 to examine an
alternative configuration. This had narrow
curved bifurcated intakes to the rear of the
nose, a large conical radome and an air intake
canal profile that was designed to minimise
pressure losses when the air reached the
compressor. The inlet duct was profiled so
that the ingested air actually received some
compression before reaching the engine, and
the radome was now fixed instead of having
to move forwards or backwards inside a nose
intake central shock cone.

The T-49 was actually the unfinished T-39
rebuilt with a completely new forward fuselage. Its conversion began in June 1958 and
the aircraft flew in January 1960. Once in the
air it displayed good flight characteristics,
with a substantial improvement in acceleration, but the test programme lasted only three
months because the T-49 was damaged in an
accident. It was repaired and received some
modifications but never flew again. This prototype has been described here because it
represents a stage in Sukhoi's move away
from nose intakes, driven in part by the
growth in size of new radar scanners. However, Sukhoi was to keep faith with the delta
wing for some time yet.

Sukhoi P-1

The first proper attempt at replacing the nose
intake came in the form of this two-seat interceptor prototype. A MAP request for a new
type was made towards the end of 1954, a
need having been identified for a new all-
weather fighter armed with guns, rocket projectiles and air-to-air missiles. Sukhoi's
studies began late in the year and initially
looked at a whole variety of weapon combinations and alternative engines, and either
one or two crew, before two related designs
were selected as the best options. The P-1
would have a single Lyul'ka AL-9 engine, the

P-2 (below) a pair of Klimov VK-11 units, and
on 19th January 1955 a directive was issued
by the USSR Council of Ministers ordering
work to proceed on both. At this stage a nose
intake along the lines of Sukhoi's Su-7/Su-9
series was in place but, as new and larger
radars were considered, this had to go to
make room for a large radome. The solution
reached during 1955 was to use circular-
shaped lateral intakes placed just above the
wing roots, while the choice of the radar itself
changed from the original Almaz-7 (Dia-
mond-7) to a Uragan set and then to the new
Pantera equipment.

The advanced development projects for
both designs were ready towards the end of
1955 and full-size mock-ups were officially
examined at the end of the year. The P-1 was
accepted and cleared for detail design and
prototype manufacture, although without
some of its planned armament, but the P-2
was now cancelled. The weapons to go were
an alternative thirty 85mm (3.35in) TRS-85
unguided rocket projectiles or thirty-two
70mm (2.76in) ARS-70 Latochka (Swallow)

folding-fin air-to-air rockets, both to be used
for hitting slow-moving heavy bombers, and
they were replaced by fifty 57mm (2.24in)
ARS-57 folding-fin rockets housed inside
compartments dotted around the fuselage
nose. Two K-7 AAMs would go under the
wings but in fact none of these weapons were
ever carried (K-8 missiles may have been
substituted at a later date).

As so often happened during this period of
enormous technical advance, the airframe
was ready well before the engine and for the
first flights a less powerful AL-7F had to be
substituted. In reheat the AL-9 was rated at
98.0kN (22,045 lb) thrust, whereas the AL-7F
gave only 87.7kN (19,730 lb). The P-1's detail
design was completed in August 1956 and the
fighter made its first flight on 12th July 1957.
However, thanks to the shortfall in power, its
performance was quite a disappointment,
and its flight-testing was halted in February
1958 after only the initial trials had been completed. The estimate performance figures
with the AL-9 in place appear to be the only
data available, but they include a top speed of

2,050km/h (1,274mph) or Mach 1.93 at 15,000m (49,213ft), a time to 15,000m of 2.7 minutes and a service ceiling of 19,500m (63,976ft).

It seems that one of the P-1's problems was that the Air Force was just not interested in having it. However, Sukhoi liked the design and to keep it going did much to promote improved versions, either with a more powerful Lyul'ka AL-11 or Tumanskiy R15-300 engine or a greater variety of weaponry, but without success. In due course the P-1 was downgraded to experimental status, and then scrapped after the whole programme was terminated on 22nd September 1958; work moved on to the T-37 heavy interceptor (Chapter Five). P-1's delta wing, scaled up from the PT-8 (Chapter Three), had a single dogtooth and was swept 60° on the leading edge, except that is for a small 'kink' towards the tip set at 55°.

It cannot be stressed enough just how many different fighter designs have been built, at least in prototype form, by the Soviet Union, and particularly during the 1950s. No other fighter design team anywhere in the world produced a series of prototypes to match Sukhoi's combined studies into intake design and wing shape (although in many cases comparative assessments would have been made in the wind tunnel). Convair in America did look at nose and side intakes through its high-speed XF-92, F-102 Delta Dagger and F-106 Delta Dart series (which also used delta wings), but for

example in England English Electric could only use tunnel data to compare the Lightning's nose intake and swept wing against versions fitted with a delta, or with a solid nose and side intakes. The cost to the Soviet Union for building so many research aeroplanes, all of it towards finding the ideal solutions for new fighter designs, must have been colossal. In the 1970s a progression to computer-aided design reduced the need for so many prototypes.

Sukhoi P-2

This was another two-seat design that, as noted, was developed in parallel with the P-1. The main difference was the switch to twin Klimov VK-11 engines, the lateral intake arrangement ensuring that each unit could have its own air supply; the intakes themselves were unchanged from the P-1, as was the size of the centre fuselage. Overall the P-2 would have been heavier, it was intended to carry more fuel than the P-1 and, unlike that aircraft, it would have guns, two NR-30 30mm cannon being housed in the wing roots (some documents, however, do refer to the P-1 having two 30mm in the lower nose). There were also underwing pylons ready to take two K-7 missiles, or twenty TRS-85 or sixteen ARS-70 rockets. The chosen radar was a new type called the Pantera but the P-2 never reached the hardware stage, the project being abandoned in late 1955 in favour of the P-1. By the way the 'P' on these projects stood for *perekhvatchik* or interceptor.

Yakovlev Yak-121 supersonic interceptor prototype.

Some mention must be made of the Klimov VK-11, the development of which began in 1954 for operation at high supersonic speeds. It is believed that this engine was expected to give about 59.8kN (13,450 lb) of dry thrust and 88.2kN (19,840 lb) in afterburner. This was a unique engine in that, in its original form, it used a compressor that combined a centrifugal section at the front followed by an axial section, but when the first examples began bench running in 1956, the 'mixed' compressor had been replaced by a short five-stage axial version. The VK-11 was also originally earmarked for the Yak-129 (Yak-28) supersonic bomber and for the Mikoyan Ye-150 and Sukhoi T-37 heavy interceptors (Chapter Five), but it was not put into production.

Yakovlev Supersonic Studies

Yakovlev Yak-121/Yak-27

Having got the Yak-25 interceptor under way, Aleksandr Yakovlev moved on to more advanced supersonic developments of the basic design. The Yak-123/Yak-26 was a prototype development for a tactical bomber that flew in 1955. It did indeed achieve supersonic flight and was followed by other bomber and reconnaissance prototypes (the bomber variants of Yakovlev's twin-jet combat aircraft family are described in the sister volume

The first prototype Yakovlev Yak-28P.

Soviet Secret Projects: Bombers since 1945). These designs were accompanied by efforts to produce a supersonic interceptor, the need for which was a growing concern, and a Council of Ministers directive was finally issued on 30th March 1955 requesting a supersonic derivative of the Yak-25. It was to be powered by two RD-9F engines and carry a Sokol-M radar and two 37mm cannon.

The resulting Yak-121 was based on the Yak-25 but had a more pointed nose that increased the length to 17.335m (56ft 10½in), a slightly greater wing area due to the addition of leading edge root extensions (although the span was unchanged), different engine nacelles and various minor changes. The first prototype flew in the early spring of 1956, as planned with less powerful RD-9AK units, but once again the desired powerplant was delayed, holding up the whole certification programme. However, despite lacking enough power, which prevented the first machine from achieving most of the specified performance figures, sufficient testing was possible to identify some necessary design changes. In due course a second directive was issued for a batch of RD-9F-powered pre-production aeroplanes to be called Yak-27. The first of these flew in July 1957 and two versions were planned, gun-only and missile-only. However, the progress made by both Mikoyan and Sukhoi with their new fighters, which offered superior performance to the Yak-27, meant that the gun-armed version would be a waste of time and so it was cancelled.

The planned service aircraft was, therefore, the Yak-27K with the Bisnovat K-8 AAM (this was one of the new weapons tested by the Yak-25K mentioned earlier). A thorough flight- and weapon-testing programme was carried out throughout 1957 and into 1958, the aircraft achieving a maximum of Mach 1.25, and as a whole it was well received. However, by now the Sukhoi Su-9 interceptor and weapon system had displayed a far superior performance, again well beyond what the Yak-27K could offer, and there was no point in taking this version of Yakovlev's aircraft any further either. The Yak-27K's service ceiling was 16,300m (53,478ft). The original Yak-121 prototype was displayed at the 1956 Tushino air show and consequently received the codename *Flashlight-C*.

Yakovlev Yak-27V

Chapter Three referred to the Mikoyan Ye-50 mixed-jet/rocket-powered fighter. Another attempt at producing a mixed-power type was this version of the basic Yak-27 fitted with a 12.7kN (2,865lb) Dooshkin S-155 rocket motor in the base of the fin. It was produced in response to yet another government order, originally made in 1955, for a single-seat high-altitude interceptor (the Ye-50 was also proposed against this document). Numerous changes of equipment included the new Almaz (Diamond) radar. The old two-seat Yak-121 airframe was converted to serve as a prototype and flew in this guise for two years from 26th April 1957. It showed promise but no production orders were forthcoming, one of the reasons being that looking after a rocket motor on a day-to-day basis was quite difficult, not least because of the toxicity of its fuel. During its trials the Yak-27V reached a maximum altitude of 23,500m (77,100ft) and a top speed of Mach 1.8.

Yakovlev Yak-28P

The next in this long line of combat aircraft was the Yak-129 tactical bomber, based on the Yak-26, which was to be powered by two Tumanskiy R-11-300 or Klimov VK-11 engines. In appearance the Yak-129 looked very similar to the older Yak-26 but in fact it was pretty much an all-new design and, with these new engines, offered a much higher speed than previous members of the family. The Tumanskiy-powered variant flew in 5th March 1958 and eventually entered service as the Yak-28 *Brewer*; it was produced in several versions.

The improved capability of this design corrected some of the weaknesses suffered by the Yak-27 and an interceptor derivative soon followed. This was the Yak-28P flown in 1960. By now bombers were having to fly at low altitudes and high speeds to try to avoid enemy defences and so the Yak-28P's task was to deal with enemy aircraft flying at low or medium altitudes, subsonically or supersonically, in all weathers, day and night. Quite a job, but to help out the interceptor would have the R-8M-1 AAM, an improved version of the Yak-27's K-8, which was now available in infra-red homing or semi-active radar homing forms, and the Koonyavskiy Oryol-D (Eagle-D) radar. It would not carry any guns. At this time the PVO was operating the Su-9 but the Yak-28P/K-8M-1 weapons system was altogether superior with both missiles and radar having a greater range than the equipment used by the Sukhoi product.

Some of the test pilots who evaluated the Yak-28P were initially against its adoption for service, declaring that the type's relatively slow maximum speed made it virtually obsolete. However, it was capable of making interceptions at low altitudes, an important factor because look-down shoot-down radars were not yet available and low flying was dangerous, requiring a pilot's full attention. Here the Yak-28P's two-man crew counted in its favour. The Yak-28P entered production in 1962 and served with the PVO for many years. In 1967 it was displayed in public at the Domodedovo air show, which ensured that it received the new Western codename of *Firebar*.

Sukhoi Su-15 Family

Sukhoi T-58 and Su-15

In the late 1950s and early 1960s the development of the USSR's combat aircraft was heavily influenced by leader Khrushchev's desire to give priority to missiles systems, both offensive and defensive, ground and air-launched – in fact his thoughts were dominated by 'missilisation'. However, that did mean that interceptors were protected a little bit more than other types against cutbacks because Khrushchev looked at them as missile-carriers; in contrast pure fighters and ground-attack types were very vulnerable. In the 1958-1959 period alone 35 aircraft projects and 21 engine types were cancelled and during 1960 Khrushchev ordered that the development of all-new aircraft should cease as an economy measure – only the modernisation of existing types was permitted. In addition, work on many new types like the Su-11, already well under way, was interrupted while some OKBs (including Lavochkin) ended their involvement with aircraft during this period.

In 1960 work began on another 'modified' interceptor when Pavel Sukhoi proposed a weapon system based on two missiles and an improved radar. The great advance that this aircraft would introduce over the Su-9 and Su-11 was the ability to make attacks from any angle, not just from the rear as in the case with the older aircraft, and at a much greater range. Work began under Sukhoi designation T-58 for a single-engine type powered by the Su-11's Lyul'ka AL-7F2. The design differed from the Su-11 through the introduction of a new forward fuselage, a larger radome and box side intakes replacing the old nose inlet; as such it was also known as the Su-11M. One solution that was rejected was to keep the Su-11-style intake but with a fixed centre cone, the inflow of air being controlled instead by air bleed flaps on the sides of the intake; this proved unsatisfactory because it gave regular and frequent compressor surg-

ing. Eventually the T-58 was given two-dimensional rectangular intakes with vertical airflow control ramps and a complex nose with the perfectly conical radome mated to a cylindrical forward fuselage: the latter being based on an arrangement first proposed for the T-37 project (Chapter Five) but rejected. The fuselage was flattened near the cockpit to allow the intakes to fit alongside.

A batch of five was ordered, including the static test specimen, and manufacture of the first prototype got under way in mid-1960. In November an official directive specified that the Volkov Vikhr-P (Whirlwind) radar should be fitted with two K-40 AAMs as the armament. The allocated service designation was Su-15, but an official directive for work to proceed never arrived, primarily because the missiles were not yet available. The prototype was due for completion in September 1961 but during mid-summer work on it was suspended. Estimated performance figures were top speed 2,650km/h (1,647mph), service ceiling approaching 20,000m (65,617ft) and range 1,900km (1,181 miles), but the Air Force was now demanding an even higher performance and something else had to be done. The prototype T-58, however, did not go to waste.

In late 1960, as an insurance, Sukhoi had started a twin-engined T-58 fitted with two 70.6kN (15,875lb) R-21F-300 turbojets mounted side-by-side in the fuselage, which gave predicted performance figures even better than before – 2,800km/h (1,740mph) max-

imum speed, up to 23,000m (75,459ft) ceiling and 2,200km (1,367 miles) range. The engine was a development of the Tumanskiy R-11F-300 that already powered the Mikoyan Ye-8, and results suggested that the twin-engined option would be the better approach. The AL-7F lacked sufficient power for this relatively heavy aeroplane (and suffered unreliability) and this contributed to the single-engine T-58 being dropped. However, the part-complete prototype was used in the manufacture of the first T-58D (the designation given to the new type) although, as usual, the planned engines were delayed (in fact, as noted already, this engine was eventually cancelled). The two T-58 variants had much in common.

To keep things moving, two 60.8kN (13,670lb) R-11F2S-300 units were fitted instead, this powerplant having previously been given a run in the T-5 prototype (Chapter Three). With all of these prototypes around, Sukhoi was often able to make use of spare airframes to help save time and money and so the back end of the T-5 prototype also found its way into the first T-58D's construction. The fuselage behind the intakes also had to be slimmed to make room for the larger supply of air required by the new engines. After a great deal of debate, it was eventually decided to mate the aircraft with the Oryol-2 radar and Bisnovat K-8M2 missile.

In the meantime Sukhoi needed to get some official support for his new interceptor, which had been started at the OKB's own risk because of the need to stay in business, but a resolution clearing the T-58 programme was not made until 5th February 1962 when work on the first prototype was nearly complete. In many ways the T-58 was a modernisation of the Su-11, which had suffered handling problems in flight because its long and heavy nose had upset the balance of the aeroplane. That aircraft's production line at Novosibirsk had been closed after a relatively small number had been built and it was replaced by the Yak-28P, Alexandr Yakovlev having used his influence to help get a decision in his aircraft's favour. Therefore, Sukhoi felt that another good product was vital because the competition for factory production space was so severe – in fact in the past some excellent Soviet aircraft designs had been dropped because there was nowhere to build them. The first T-58 finally left the runway on 30th May 1962 and the second, complete with its radar, followed it into the air nearly a year later.

The first pre-production Su-15 displays its large radome, delta wing and air-to-air missiles.

Soviet Secret Projects: Fighters Since 1945

This view of the upgraded Su-15TM shows the conical radome well, but the cranked delta wing is less obvious. This is a late-production aircraft with four missile pylons and two drop tanks.

Sukhoi Su-19M (1972). The ogival wings are clearly shown.

As an interceptor the T-58 proved far superior to the Yak-28P in terms of flight speed and performance; its ceiling for example was 18,500m (60,696ft) when the Yak-28P's was 16,000m (52,493ft), although the latter's range was better. The T-58's superiority was clear to all and eventually the Yak-28P's Novosibirsk line was closed with production switching to the T-58 (which must have pleased Pavel Sukhoi no end). In service the aircraft was still called the Su-15 and the West coded it *Flagon* (the service designation had nothing to do with the first Su-15 described above). After joining the PVO in 1967 the type gradually replaced the Yak-28P and the Su-9 and Su-11. Several versions were built including the upgraded Su-15TM, which had a cranked delta wing rather then the original pure delta. The original Su-15 was thought to have a wing that was undersized for its weight and the Su-15TM corrected the fault by introducing tip extensions. The new mark began its official test programme in September 1970. By then the improved Taifun-M radar had been put in place, housed in an ogival radome instead of the old conical form. Su-15 withdrawals did not begin until the early 1980s when the Mikoyan MiG-31 and Sukhoi Su-27 became available.

Sukhoi T-59
One of a series of parallel projects intended to study alternative configurations to the early T-58, this design was a development of the very fast T-37 interceptor covered in the next chapter. The work undertaken on the T-59 was brief, but the design had lateral intakes along the lines of the T-49 prototype and was intended to carry Sukhoi's own K-9-51 missile plus a TsP radar. The T-59 would also have acted as a testbed for the radar.

Sukhoi T-60
Another parallel study, this was really a version of the twin-engined T-58 but with oblique box intakes along the sides of the fuselage level with the cockpit. The larger inlets did not need a waisted fuselage to supply sufficient

air to their R21F-300 engines, but in most other respects the designs were quite similar.

Sukhoi Su-19 and Su-19M
Rounding off the Su-15 story, during the first half of the 1970s attempts were made to produce an advanced interceptor development based on the original aircraft. The first of these was called the T-58PS and was studied in 1972 and 1973. The fuselage was given new larger wings of ogival form based on those fitted to the first T-10 (Su-27) prototype (Chapter Nine), which offered much improved aerodynamics, and tunnel testing at TsAGI confirmed that both the aircraft's performance and agility would be far superior to the basic Su-15. This meant that the new aircraft would be more manoeuvrable in a dogfight and to back this

Model of the Sukhoi Su-19M. John Hall

up the new wing added two more hardpoints for short-range missiles. This made eight in all, six under the wings and two under the fuselage, and there was also a built-in cannon on the fuselage centreline. During the later stages of the project's life the official correspondence between the Design Bureau and its customers referred to it as the Su-19.

Next came the Su-19M with a new R67-300 engine added to the re-winged airframe, which offered more power and better economy and thus improved range and better acceleration; it was also a smaller engine which made space for more fuel. Sukhoi's official proposal stated that the Su-19 with just the new wings could fly during the last three months of 1973 while the full Su-19M would get into the air in the first quarter of 1975. However, what this aircraft lacked was a look-down/shoot-down radar, which was becoming a key element in defeating bombers flying at very low level, and so no interest in the Su-19M was forthcoming from the Air Force or other official circles. That brought the end of the Su-19 series but one final proposal had a Poorga (Snowstorm)

look-down/shoot-down radar inside the basic Su-15TM airframe, four K-50 AAMs and a 109.9kN (24,725 lb) reheated thrust Lyul'ka AL-21F-3 (which also powered Sukhoi's Su-24 *Fencer* strike aircraft). The whole package was called the Su-15M and this did generate some support from the PVO, but MAP was not interested and so all studies into upgrading the Su-15 finally came to a close.

Yakovlev Developments

Yakovlev Yak-28-64

When the Su-15 first entered production at Novosibirsk it joined the Yak-28P line, which meant that one factory was producing two entirely different aircraft that were to perform the same tasks, and they were equipped with the same engine (the R-11F2-300) and radar (the Oryol-D). The superior performance of the Su-15, with two engines housed neatly inside the fuselage, over the Yak-28P and its underwing nacelles forced the Yakovlev OKB to take a good look at its rival. One benefit available to Aleksandr Yakovlev was that

his staff could inspect the Su-15 with ease because both types were under the same roof, so with some haste he sent his son Sergey to Novosibirsk to take a look at, in particular, the drawings of the Su-15's air intakes. The prototype Mikoyan MiG-25 (Chapter Eight) was also examined in detail.

The result was a radical development of the Yak-28 exhibiting an Su-15-style configuration with buried engines and two-dimensional lateral box intakes. The centre and rear fuselage was much wider than on the Yak-28P and the intake ramps and boundary layer splitter plates were pretty well direct copies of Su-15 practice. For the rest of the design, the original Yak-28 wings and planform were retained (minus the nacelles) and the forward fuselage, tail and undercarriage were unchanged. The wing outer panels were reduced to keep the span roughly the same and two strakes were placed beneath the rear fuselage to help the directional stability. There were two pylons on each wing, the inner carrying R-8M-1 AAMs while the outer pair were loaded with dummy examples of a 'future' short-range 'dogfight' weapon currently under development. It was hoped that these changes would give a better maximum speed and ceiling and a product that, overall, was superior to the Su-15. An additional worry was that the original Yak-28 layout had reached the end of its development and the OKB required another product to gain new orders.

This work began in 1964 and so the fighter received the provisional designation Yak-28-64. It was an impressive beast but also looked rather overweight. The prototype, converted from a standard Yak-28P, was rolled out in 1966 and is believed to have flown on 5th November. In the air it was not a success. The first series of flight tests revealed a pretty poor performance, worse even than the old Yak-28P with its underwing power units. Its handling was inadequate and the aircraft suffered aileron reversal. The latter had also been a feature of the original Yak-28's handling but it had been hoped that removing the underwing nacelles, which made room for larger twin ailerons, would have cured this problem and also increased the maximum speed at low level; it did not.

Throughout this and the previous chapter great stress has been made in regard to Sukhoi's studies into air intake design, and this had now paid off. Years of experience went into the Su-15's adjustable lateral intakes whereas Yakovlev really had to start from scratch on the Yak-28-64, having had no previous experience with this arrangement. The resulting design was produced in haste, which meant there was insufficient time to

Top and centre: **The Yakovlev Yak-28-64. It is a pity that there appears to be no air-to-air views of this very impressive-looking machine.**

Bottom: **Model of the Yak-28-64.** John Hall

find, let alone deal with, certain weaknesses, and it proved to be flawed. Having had a lot of money spent on it, the Yak-28-64 was quickly abandoned, much to Yakovlev's disappointment. It is believed that some drawings were produced in readiness to build a second prototype.

Yakovlev Yak-35MV

Although earlier in timescale than the Yak-28-64, this project initially had little to do with any rival products from the other OKBs. In 1958 Yakovlev began working on a low-altitude fighter and interceptor project designated Yak-35MV (MV for *malovysotnyy* or low-altitude). It was to be powered by a single RD11-300 turbojet (the early designation for the better known R11-300) and was intended to destroy enemy aircraft flying at altitudes between 200m (656ft) and 10,000m (32,800ft). This appears to have been one of the first occasions that the need to deal with targets flying at low-level had been acknowledged and to help with this task the Yak-35MV was to be part of the Vozdukh-1 (Air-1) ground-controlled intercept system. A specification for the interceptor was covered by a Government directive of 4th June 1958 which also stipulated that the type was to be submitted for its state acceptance trials during the last quarter of 1961. It was expected to be capable of 1,300km/h (808mph) at sea level and 1,550km/h (963mph) at 10,000m and carry 6,000kg (13,228lb) of fuel, two 18km (11.2 miles)-range K-35 AAMs and a Pantera-2 radar with a detection range of 30km (18.6 miles).

The fighter would also serve as a basis for a tactical fighter-bomber that was expected to incorporate a number of features for enhancing its chances of surviving heavy anti-aircraft artillery and small arms fire. The Yak-35MV fighter-bomber was a single-seat aircraft with a take-off weight of 15,000kg (33,069lb), which included 2,000kg (4,409lb) of weapons. Its design performance included a maximum speed (with external bombs) of 1,300km/h to 1,350km/h (808mph to 839mph) at sea level and 1,500km/h to 1,600km/h (932mph to 994mph) at 10,000m (32,808ft), and a service ceiling of 10,000m (32,808ft); power would come from two Tumanskiy R11A-300 engines. The fighter-bomber version came into being because Air Force officials felt that some new types planned for the Service's frontline strike units, the Yak-26, Tupolev

Tu-98 and Ilyushin Il-54 bombers (the former a development of the Yak-25), would be inadequate for the job they were required to do. As a result the Yakovlev OKB began its fighter-bomber 35MV, as well as continuing its Yak-25/26 family, which included a reconnaissance development called the Yak-32.

However, by now Sukhoi had started to convert the original Su-7 tactical fighter (Chapter Three) into the Su-7B fighter-bomber and official assessments indicated that the Yak-35MV bomber was believed to be little different from the 'almost ready' Su-7B. By

1958 plans had been made to put an enormous number of new strike aircraft into service but their cost, together with Khrushchev's growing interest in missiles, brought the abandonment of the greater part of this programme. One design to survive was the Yak-28 strike aircraft, from which the Yak-28P (above) was developed, and this eventually entered service along with the Su-7B. The 35MV scored over the Su-7B in having two engines and, in consequence, better survivability during low-level operations, but it also suffered from delays to the development of its engines. The

Su-7B and Yak-28 were considered to be stopgaps but the Yak-35MV bomber and the Yak-32 were eventually cancelled because of budget cuts and delayed engines.

All of these factors did not help the single-engine Yak-35MV interceptor either, which again suffered from a lack of funding. In 1960 this programme was also terminated. Drawings for the two versions of the Yak-35MV have not been released but the fact that there were both single and twin-engined versions suggests that their powerplants were housed within the fuselage.

Radar-Equipped Fighters – Data / Estimated Data

Project	Span m (ft in)	Length m (ft in)	Gross Wing Area m² (ft²)	All-Up-Weight kg (lb)	Engine kN (lb)	Max Speed / Height km/h (mph) / m (ft)	Armament
Lavochkin La-200 1st Prototype (flown)	12.96 (42 6)	16.59 (54 5)	40.02 (430.3)	10,375 (22,873)	2 x VK-1 26.4 (5,950)	964 (599) at S/L, 1,090 (677) at 3,500 (11,483)	3 x 37mm cannon
Lavochkin La-200B (flown)	12.96 (42 6)	17.325 (56 10)	40.02 (430.3)	11,560 (25,485)	2 x VK-1A 30.4 (6,835)	1,030 (640) at 5,000 (16,404)	3 x 37mm cannon
Mikoyan I-320 (R-1) (flown)	14.2 (46 7)	15.775 (51 9)	41.2 (443.0)	10,265 (22,630)	2 x RD-45F 22.2 (5,005)	1,040 (646) at S/L, 994 (618) at 10,000 (32,808)	2 x 37mm cannon
Sukhoi Su-15 (1st) (flown)	12.87 (42 2.5)	15.44 (50 8)	36.0 (387.1)	10,437 (23,009)	2 x RD-45 21.6 (4,850)	1,032 (641) at S/L, 1,045 (649) at 5,000 (16,404)	2 x 37mm cannon
Yakovlev Yak-50 (flown)	8.01 (26 3)	11.18 (36 8)	16.0 (172.0)	4,100 (9,039) 1st prot, 4,155 (9,160) 2nd prot	1 x VK-1 26.4 (5,950)	1,170 (727) at S/L, 1,135 (705) at 5,000 (16,404)	2 x 23mm cannon
Yakovlev Yak-60	?	?	23.0 (247.3)	4,000 (8,818)	1 x VK-1 26.4 (5,950)	Mach 0.9+	Cannon
Yakovlev Yak-120 1st Prototype (flown)	10.964 (35 11.5)	15.665 (51 4.5)	28.94 (311.2)	9,450 (20,833) with tanks	2 x AM-5 19.6 (4,410)	1,140 (709) at 4,000 (13,123), 1,090 (677) at 5,000 (16,404)	2 x 37mm cannon
Yakovlev Yak-25M Early Production (flown)	10.964 (35 11.5)	15.665 (51 4.5)	28.94 (311.2)	10,045 (22,145) with tanks	2 x RD-5A 19.6 (4,410)	?	2 x 37mm cannon
Lavochkin La-190 (flown)	9.9 (32 6)	16.35 (53 7.5)	38.93 (418.6)	9,257 (20,408)	1 x AL-5 51.0 (11,465)	1,190 (740) at 5,000 (16,404)	2 x 37mm cannon
Mikoyan Ye-8 (flown)	7.154 (23 5.5)	16.24 (53 3.5) including pitot	23.13 (248.7)	8,200 (18,078)	1 x R21F-300 46.0 (10,360), 70.6 (15,875)	2,230 (1,386) at height	2 x K-23 AAM (carried 2 x K-13)
Sukhoi P-1 (flown)	9.816 (32 2.5)	21.83 (71 7.5)	44.0 (473.1)	10,600 (23,369) normal, 11,550 (25,463) max	1 x AL-7F 67.1 (15,100), 87.7 (19,730) reheat	2,050 (1,274) at 15,000 (49,213) (Estimated with AL-9 engine)	2 x K-7S AAM, 50 x 57mm RP
Yakovlev Yak-27K (flown)	10.964 (35 11.5)	17.335 (56 10.5)	?	10,680 (23,545)	2 x RD-9F 26.9 (6,060), 37.2 (8,375) reheat	1,270 (789) at height	2 x K-8 AAM
Yakovlev Yak-28P (flown)	11.64 (38 2)	20.55 (67 5)	35.25 (379.0)	16,065 (35,417) normal	2 x R-11AF2-300 38.7 (8,710), 59.8 (13,450) reheat	1,840 (1,144) at 12,000 (39,370)	2 x R-8M-1 AAM
Sukhoi Su-15 (flown)	8.616 (28 3)	20.54 (67 4.5) without pitot	34.56 (371.6)	16,520 (36,420) normal	2 x R11F2S-300 38.2 (8,600), 60.8 (13,670) reheat	1,200 (746) at S/L, 2,230 (1,386) at 15,000 (49,213)	2 x R-98/T or R-8MR/MT AAM, 23mm cannon in underfuselage pods
Yakovlev Yak-28-64 (flown)	c11.8 (38 8)	c22.18 (72 9) with pitot	?	?	2 x R-11AF2-300 38.7 (8,710), 59.8 (13,450) reheat	Unknown but less than Yak-28P	2 x R-8M-1 AAM, 2 x small AAM

Fighters With Radar

Chapter Five

Heavyweights

The PVO air defence force's assignment of building a chain of radars along its frontiers and fielding new patrol interceptors to go with it was an enormous venture which received the highest priority. As an interceptor the Yakovlev Yak-25 did a good job but within a few years new advances in bomber capability meant that the type could no longer cope with fast high-flying targets. The solution was to develop a new supersonic interceptor armed with the most advanced air-to-air missiles. Some of the types intended to fulfil this general demand were discussed in Chapters Three and Four but another set of altogether more potent aeroplanes was also produced. All of the Soviet 'fighter makers' had a go at developing advanced aerial intercept weapons systems. Despite leader Nikita Khrushchev's love of missiles and the fact that a number of fighter projects were terminated because of this, during the late 1950s and early 1960s the output of Soviet fighter

prototypes seems to have been relatively undiminished.

Last Lavochkin

Lavochkin 'Aircraft 250'

The first people to propose a combined weapons system, in 1953, were probably Semyon Lavochkin and Viktor V Tikhomeerov. The former was the leader of the famous fighter OKB while Tikhomeerov was the director of NII-17, a Ministry of Aircraft Industry establishment working on avionics. Lavochkin was willing to design both the fighter itself and its principal AAM weapons while Tikhomeerov would be responsible for the aircraft's fire control radar. Their proposal won official support and on 20th November 1953 the Council of Ministers formalised its approval with a directive ordering full development of the K-15 aerial intercept weapons system. This embraced

A 1954 artist's impression of how the Lavochkin '250', or La-250 as it was also known, would have looked in flight.

four components – the '250' long-range interceptor, K-15U radar, 'Izdeliye 275' missiles and their ground-controlled intercept equipment: '15' referred to the missile's maximum range of 15km (9.3 miles).

Lavochkin had taken on a massive task, which involved not only the creation of an interceptor of a totally new class but also the missiles to go with it – an area in which the fighter Design Bureau had no prior experience. Every aspect of the design was new and, just to complicate the issue, the development schedule was also extremely tough. Initial sketches were made in the summer of 1953 and were based on a design that looked very similar to the '200' described in Chapter Four, but with a horizontal tail of unusual rhomboid shape. Power came from two

The Lavochkin '250A' demonstrates the original nose. It also carries two missiles, one beneath each wing.

Lyul'ka AL-7 engines that flanked the centre fuselage, there was a large radar in the nose, two AAMs were carried and the two crew sat side-by-side. All-up weight was 21,000kg (46,296 lb) and the design's estimated performance figures included a top speed of 1,600km/h (994mph) at 12,000m (39,370ft), a time to 12,000m of between 2.5 and 3.0 minutes and a ceiling of 16,000m (52,493ft).

This layout was eventually dropped and the resulting 'Aircraft 250' was designed around two Klimov VK-9 afterburning turbojets rated at 117.6kN (26,455 lb) each, the latest product from the famous engine OKB. According to the covering operational requirement the '250' was expected to deal with targets flying at up to 20,000m (65,617ft) altitude, 1,250km/h (777mph) speed and within a radius of 500km (311 miles). The interceptor would be guided into the vicinity of the target by a Vozdookh-1 (Air-1) ground station and a Lazoor (Prussian Blue) command link system, after which it would then acquire the target with its own radar. Two semi-automatic radar homing *Izdeliye* 275 AAMs (which

needed the target to be illuminated all the way from launch to impact) would be carried in tandem semi-recessed in the fuselage.

By July 1954 the general design of the '250' was nearly complete and a full-scale wooden mock-up was built; its profile ensured that the interceptor was nicknamed Anaconda. At first the work on the K-15 weapons system moved ahead with speed but it soon became apparent that development of some of the system's primary elements, particularly its radar and engines, was late. Then, in early 1955 when the construction of the first prototype was already under way, the VK-9 was cancelled. No existing alternative would fit the aircraft's size and weight and so, to meet the required performance figures, Lavochkin chose the Lyul'ka AL-7F as a replacement. The AL-7F was far less powerful than the VK-9, being rated at 73.5kN (16,535 lb) dry and 98.0kN (22,045 lb) with reheat, and so in effect the '250' had to be redesigned from scratch around a different powerplant. In-house the aircraft was redesignated 'Aircraft 250A' and, overall, was smaller and lighter.

The '275' missiles were also redesigned and moved to underwing pylons.

The '250A' advanced development project with slightly uprated AL-7F-1 engines was completed in October 1955 and approved in early 1956. Overall the interceptor's performance was slightly reduced while the flight parameters for the targets had been revised to 19,500m (63,976ft) altitude and 1,200km/h (746mph) speed. In the meantime the original '250' prototype was completed on 16th June 1956, work on it having continued because Lavochkin had decided to use the aircraft for preliminary flight tests and as a systems testbed. The AL-7F-1 was also not yet available and so the '250' was fitted with standard AL-7Fs rated at 67.6kN (15,210 lb). Preflight ground testing went well and on 16th July the '250' made its first 'flight', except that two seconds after getting airborne it banked several degrees to starboard and then started pitching and rolling. The pilot opted for an immediate landing, after which the aircraft caught fire and was written off. The cause was the control system, which had to be redesigned.

When completed in the spring of 1957 the first '250A' had the modified control system and also AL-7Fs because the AL-7F-1 was still unavailable. There were numerous other changes including intakes placed further forward and differences around the jetpipe, which to a degree had altered the interceptor's appearance. The first '250A' flew on 12th July 1957 but was to become another loss, written off in a landing accident on 28th November. This time pilot visibility was the problem and so the third aircraft (that is, the second '250A') had the nose ahead of the windshield angled down by 6° (right from the mock-up stage several pilots had declared that the long nose would give problems with external vision). This aircraft flew in July 1958, still powered by AL-7Fs (and without working afterburners), which meant that the flight envelope could not be explored in full.

The first three '250' and '250A' prototypes made their maiden flights at roughly twelve-month intervals and the long delays being experienced with the engines put the programme way behind schedule. A K-15M radar was finally fitted to the fourth prototype while the fifth machine had a complete weapons and avionics suite. At last, in May 1958 all of

The third '250A' with its modified nose.

Soviet Secret Projects: Fighters Since 1945

the components of the K-15 weapons system – aircraft, missiles and radar – were ready for testing, but it was too late. The delays had left their mark and the Commander-in-Chief of the Soviet Air Force, Marshal Konstantin A Vershinin, voiced criticism of the aircraft's time schedule and added that the whole system was becoming obsolete. As a result, in July 1959 the government issued a directive cancelling the K-15 weapons system as a whole, including the '275' missile. By now work was also progressing on the Su-9 (Chapter Three) and the competing Tu-28-80 (below) which would more than fill the gap left by the disappearance of the '250A'.

Even allowing for the development problems and the fact it did not receive the engines it needed, Lavochkin's interceptor exhibited a disappointing performance. Top speed was just 1,080km/h (671mph) instead of the desired 1,610km/h (1,001mph), service ceiling 13,300m (43,635ft) instead of around 17,000m (55,774ft) and time required to reach 12,000m (39,370ft) 5.4 minutes instead of around half that. As an Aircraft Design Bureau this was Lavochkin's swan song because the Soviet government now decided that the OKB should switch to being a specialist missile manufacturer – it was never again to be involved with fighter design although it did get involved with spacecraft. However, the '250'/K-15 design effort was not wasted because much of the data and experience gained with it was transferred to the Tupolev OKB to help with the 'Aircraft 128' heavy interceptor. Only one '250A' survives today which is held in the Monino Museum and no Western codename was allocated to the type.

To round off Lavochkin's contribution to this book, it is worth making a brief reference to the OKB's last aircraft project, which was actually a bomber. When the companion *Soviet Secret Projects: Bombers since 1945* was written it was firmly believed that Lavochkin had never worked in detail on a jet bomber, but new information has shown that this statement was incorrect. 'Aircraft 325' was a supersonic long-range bomber project and its development was officially approved in March 1956. Initially it was to be a single-seater powered by a ramjet engine and would carry 2,300kg (5,071 lb) of weapons at a top speed of 3,000km/h (1,865mph). 'Aircraft 325' would have a range of 4,000km (2,486 miles) and was to fly over its target at heights approaching 20,000m (65,617ft); in late November 1956 this figure was increased to 25,000m (82,021ft) after the ramjet was

Model of the Tupolev '128' in its ultimate project configuration.

replaced by a VK-15 engine. Work on the aircraft was slow, not least through the OKB's preoccupation with the problems of the '250', but in September 1957 the Bureau approached the VVS with a suggestion to make the '325' a pilotless aircraft. There would now be two versions, a bomber and a reconnaissance type with the latter launched from a mobile ramp. However, work on 'Aircraft 325' was abandoned in July 1958.

Tupolev Interceptor Project

Tupolev 'Aircraft 128' and Tu-28

The Tupolev Design Bureau had traditionally dealt with bombers, airliners and large aircraft in general. However, in 1957 PVO Command turned directly to Tupolev with a proposal that the OKB should produce, as a matter of urgency, a missile-armed interceptor based on the 'Aircraft 98' experimental supersonic tactical bomber first flown in September 1956. The proposal was accepted and by June 1957 the Bureau's Technical Department under S M Yeger had begun studies under the designation 'Aircraft 128'; the new type was initially given the official designation Tu-28. For over a year the work was classed as 'unofficial' because the Soviet Council of Ministers resolution giving an official go-ahead was not released until 4th July 1958. The system as a whole was known as the Tu-28-80 and was made up of the aircraft itself, Bisnovat K-80 air-to-air missile with alternative radar-guidance or heat-seeking homing heads, Smerch radar and the Voz-dukh-1 intercept guidance system. Tu-28 would have a top speed of 1,700km/h to 1,800km/h (1,057mph to 1,119mph) and had to be able to intercept subsonic and super-

The '128' prototype's flypast at the Tushino air show on 9th July 1961. The missiles were dummies.

sonic targets at altitudes up to 21,000m (68,898ft). The entire weapon complex was to be ready for its factory tests in the first quarter of 1960 and for state trials towards the end of that year.

The decision to base the Tu-28 on the comparatively unmanoeuvrable '98' bomber was fine because, unlike many other interceptors, the new aircraft was not required to be particularly manoeuvrable or engage its target at the same altitude. The K-80's great range

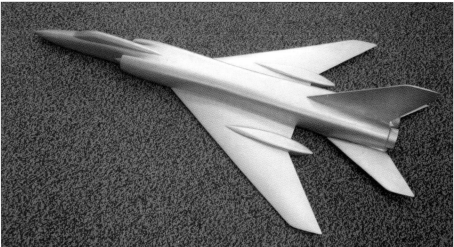

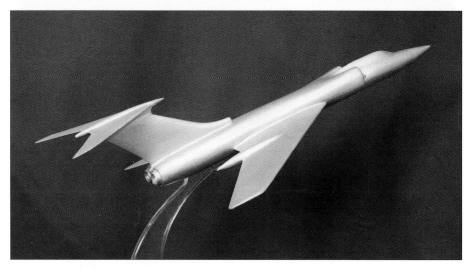

The first '128' prototype photographed in a typical Russian winter landscape in late 1961.

Models of two Tu-28 developments. One has a different fin, the other a T-tail.

actually meant that it would be possible to engage targets flying at heights well above the interceptor, so the most important phase of manoeuvre onto the target was transferred from the interceptor to the missile. Nevertheless the aerodynamics and structure of the '98' still had to be improved substantially and area ruling was applied to the airframe to enhance its aerodynamic efficiency at high subsonic cruising speeds. Power would be supplied by two Lyul'ka AL-7F-1s, although these were later replaced by a development called the AL-7F-2.

The first '128' prototype was completed in the summer of 1960 and made its maiden flight on 18th March 1961; after its display at Tushino in July the West named the aircraft *Fiddler*. In autumn 1962 the first successful interception and destruction of a target drone was achieved and in late 1963 the whole system received the service designation Tu-128S-4 – 'Aircraft 128' with the R-4 missile. State trials were successfully concluded in July 1964 and on 30th April 1965 the system was accepted for service with the PVO. Other versions were produced including the Tu-128M, an updated aircraft first flown on 24th September 1970 that had more capability for intercepting targets at low altitude. The airframe formed part of the new Tu-128S-4M system embracing the R-4TM missile and

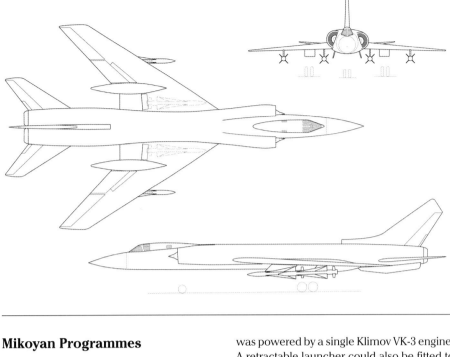

Smerch-M radar. Two prototypes were built from stock Tu-128s but the production run was made entirely of in-service aircraft upgraded to the new standard. Both service variants had a service ceiling of 15,600m (51,181ft).

Earlier, in 1962 and 1963, the OKB had worked on the Tu-28A interceptor project powered by two Dobrynin VD-19 engines, which required larger inlets and ducting plus a slightly wider rear fuselage; splayed ventral fins were also fitted. Other changes considered here included a longer-range radar and missiles (Smerch-A and K-80M) and a longer nose to house the radar. The new engines offered a potential top speed of over 2,000km/h (1,243mph), but the Tu-28A and its Tu-28A-80 system were eventually rejected. Another discard was the Tu-28A-100 that would have had the Groza-100 fire control radar and K-100 missiles. There were also proposals for a multi-role aircraft armed with unguided and guided missiles, bombs and cannon, and the '128B' tactical bomber variant, but none of these were built.

The Tu-128S-4M served with PVO until the second half of the 1980s but after the end of the Cold War most examples were withdrawn and scrapped. As an interceptor the Tu-128 was a huge machine and reflected a tendency of the time in that combat performance could be improved by carrying more capable weapons, but at the expense of the aircraft's flight capabilities. It was also a very heavy aircraft and lacked the performance of other interceptors. Part of the problem was that Tupolev was not an established fighter producer and some design practices used by the OKB for large aeroplanes were carried over into the Tu-128. Nevertheless, the *Fiddler* was a successful aircraft.

Mikoyan Programmes

During the 1950s the Soviet Union explored several ways of protecting its homeland against attack by enemy aircraft, some of which have been described already. One result of the research was the Uragan-1 (Hurricane-1) air defence system designed specifically to cover the USSR's industrial heartland and the airframe elements of later versions of this system were to become objectives of the Mikoyan and Sukhoi Design Bureaux.

Mikoyan I-3 or I-380
This story begins on 3rd June 1953 when a new Council of Ministers decree ordered Mikoyan to build an experimental tactical fighter. The resulting design had three NR-30 30mm cannon fitted in the wing roots and

was powered by a single Klimov VK-3 engine. A retractable launcher could also be fitted to hold sixteen 57mm folding-fin rockets while the underwing pylons could carry two 190mm (7.48in) or 212mm (8.35in) high-velocity rockets, two 250kg (551lb) bombs or two standard 760-litre (167-gal) drop tanks. Estimated service ceiling was 18,800m (61,680ft). The design was called the I-3 or I-380 and its advanced development project was completed in March 1954. The covering decree required the first prototype to commence its state trials during the first three months of 1956. Although the I-3 prototype was built it was, however, never to fly in its original form because a flight-standard VK-3 was never delivered.

Mikoyan I-3P or I-410
Running in parallel with the I-3, and covered by the same Council of Ministers directive, was a day interceptor project called the I-3P or I-410 fitted with an Almaz (Diamond) radar. This would also be powered by a VK-3 and had to be ready for its state trials in May 1955. Prototype construction began in August 1954, by the end of January 1955 it was around 42% complete and the airframe was finished before the end of the year, but once again no flight-cleared engine was to become available and so the I-3P never flew in its original form. However, this work was not wasted because on 28th March 1956 a new directive was released ordering Mikoyan to covert the

Tupolev Tu-128M – note the different fin shape.

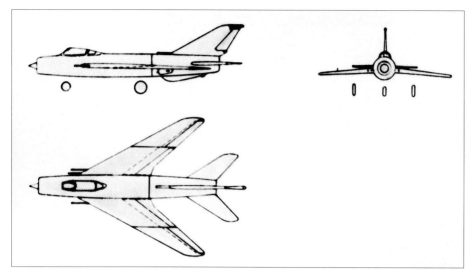

airframe to take a Lyul'ka AL-7F engine plus the radar equipment used by the Uragan-1 system.

Mikoyan I-3U or I-420

A separate programme to produce an interceptor fitted with Uragan equipment was under way in 1954 covered by a directive issued on the 2nd March. Design work began in September and the I-3U (U for Uragan) was to be submitted for state testing during the third quarter of 1956. Initially the aircraft was known as the I-420 and it eventually replaced the I-3P in the overall programme. The second unfinished I-3 airframe was available for conversion and this became the prototype I-3U, being substantially complete by the start of 1956. It was armed with just two 30mm cannon, one in each wing root. However, a flight-standard VK-3 engine was not delivered until July and then, after the flight test programme was under way, it had to be regularly returned to the manufacturer for modifications. There were also problems with the airframe's hydraulic actuators and the result of these troubles was that by 1st January 1958 only 43% of the flying programme had been finished. The I-3U achieved a service ceiling of 18,000m (59,055ft) and took 2.4 minutes to reach 15,000m (49,213ft) but the Council of Ministers brought the programme to an end on 4th June 1958 and the aircraft never flew again. Some documents have apparently called this aircraft the I-5.

Mikoyan I-7U and I-7K

Returning to the earlier line of development, the I-3P re-engined with the AL-7F powerplant was redesignated I-7. In fact two versions were planned, the I-7U armed with cannon in the wing roots and unguided 57mm rockets in the inner wings and the I-7K, which introduced two Grooshin K-6 missiles instead of the guns. The respective advanced development project documents were completed on 23rd July and 28th August 1956, the I-7K having a different nose because of the Almaz-3 radar plus fairings over the gun ports. Converted from the I-3P, the I-7U flew on 22nd April 1957 and quickly revealed that, compared to the former's fine performance,

Top: **Mikoyan I-380/I-3 (1953/54).**

Centre and bottom: **Mikoyan I-380/I-3 impressions taken from the project documents. The first picture has the aircraft fitted with a sliding canopy, but this was later replaced by the forward-hinged canopy illustrated in the second shot. There are other subtle changes to the intake cone and the wing fences while the first view shows a gun in the lower fuselage.**

Soviet Secret Projects: Fighters Since 1945

Above: **Mikoyan I-3U (I-420).**

Right: **Artist's impression of the I-7U.**

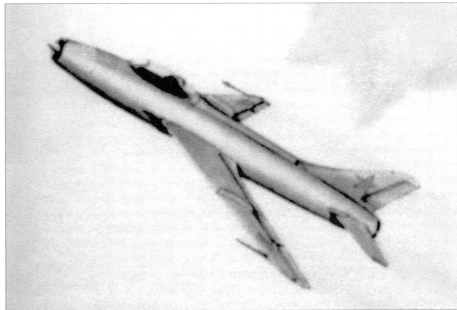

the new type was mediocre – top speed was just 1,420km/h (883mph) when the requirement asked for 2,300km/h (1,429mph). The key was the engine, which was a pre-production unit giving nowhere near enough thrust, just 86.2kN (19,400 lb) with reheat: the specified figure was 90.2kN (20,300 lb). The I-7U did achieve a service ceiling of 19,100m (62,664ft) but Artyom Mikoyan himself ended its flight test programme on 12th February 1958 and the airframe returned to the shops to be converted into another new type, the I-75 below. The I-7K prototype with twin radomes was never completed.

–

Compared to the massive Lavochkin '250' and Tu-28 discussed earlier, Mikoyan's first essays into 'heavy' interceptors had been altogether smaller aeroplanes (although they were bigger than the MiG-21). The next series was also relatively small in size but in terms of speed they were giants – even with its intended engines the '250' could never have matched the speed of the aeroplanes that follow. To go with them was the Uragan-5 sys-

tem. Begun in 1954 Uragan-5 became the USSR's first integrated automatic aerial intercept weapons system and in February 1955 the Council of Ministers ordered that the next interceptors from both Mikoyan and Sukhoi should be tied into it, together with their missiles. Mikoyan was told that it had to produce five prototype interceptors with different

weapons – two with K-7S missiles, another with the K-6V and two more with elevating cannons as a backup in case of failures in the missile's development. Together with the associated engines and avionics, this was a very complex project and the numbers of prototypes, their weapons and their development schedule was subsequently changed

Top: **The prototype I-7U. Note the cannon housed in the wing root.**

Centre: **Mikoyan I-7K with two packs of folding fin rockets in each inner wing and underwing air-to-air missiles (8.56).**

Bottom: **Prototype Mikoyan I-75.**

several times. The results were the Ye-150 series and Sukhoi's T-37, 'Ye' signifying *yedinitsa* or 'one-off'.

The potential targets were enemy aircraft flying at speeds of up to 2,500km/h (1,554mph) and heights up to 20,000m (65,617ft). Much of the Uragan-5 programme gradually fell behind schedule and in due course both Mikoyan and Sukhoi modified their aircraft to form part of the newer Dal' ground-control system, the interceptors now armed with Mikoyan's own K-9 missile designed specifically for the Ye-150 family. The respective systems were renamed the Mikoyan Ye-152-9 and Sukhoi T-3A-9 and in 1961 the former was amended to work with the Vozdookjh-1 (Air-1) automatic ground-control guidance system (as the Ye-152-9-V). However, politics and changes in the organisation of the equipment manufacturers involved brought the closure of both programmes, in 1962 and 1960 respectively.

Mikoyan I-75 and I-75F

Like its I-3 and I-7 predecessors, the I-75 featured a swept wing. This was not intended to be one of the new prototypes but was the older I-7U converted to serve as a testbed for Uragan-5's missiles (in this case two Bisnovat K-8s) and radars. The incomplete I-7K airframe was turned into the follow-on I-75F with a more powerful AL-7F-1 rather than the I-75's AL-7F, the two rebuilds introducing a much-modified nose with three-shock centrebody. The specified versions of the Lyul'ka engine were intended for Mach 2.1+ operation above 20,000m (65,617ft) but, yet again, the powerplant was late and proved to be unreliable. However, the first I-75 flew on 28th April 1958 without its radar and showed an impressive performance, including a ceiling of 18,700m (61,352ft) without weapons and 16,000m (52,493ft) with both AAMs aboard, when the top speed was 1,670km/h (1,038mph) at 12,400m (40,682ft). With the radar added it proved to be a pretty good interceptor and became a competitor to Sukhoi's Su-9 (Chapter Three). However, that aircraft's development was far more advanced and there was never much chance of an I-75-type entering production, so flight-testing ended on 11th May 1959.

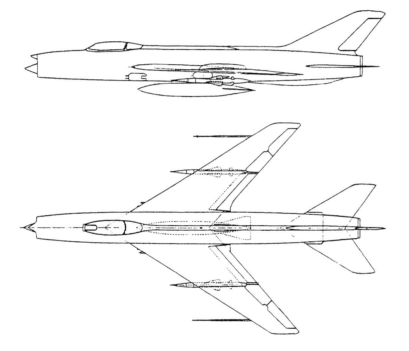

Mikoyan Ye-150

The first new Mikoyan prototype to be produced in response to the Council of Ministers directive of 26th February 1955, this was an awesome-looking machine. Go-ahead was covered by another document dated 28th March 1956 when four examples were planned – two with K-6V missiles, one with unguided rockets and the last with moving cannon (the Ye-151). After the reshuffles noted earlier, a programme of five aircraft was settled – one with K-6Vs, two with K-7S missiles and two with the guns. Ye-150 had a delta wing, a tube fuselage and a massive intake and shock cone. Achieving speeds well in excess of Mach 2.5 required powerful engines, which in this case was a Tumanskiy R15-300, a single-shaft engine with five compressor stages and a single stage turbine plus three-position variable nozzle.

The Ye-150 advanced development project was finished in the summer of 1957 and, although completed in September 1958, the prototype's first flight was held up for an incredible amount of time by ground testing and delays to the engine's preliminary test programme. As a result the first Ye-150 did not fly until 8th July 1960, after which further engine and development troubles initially reduced the number of test flights to be completed. In due course, however, it became clear that this was a very impressive beast with a superior performance to other fighters and interceptors. Service ceiling was 23,250m (76,280ft) and the time to 15,000m (49,213ft) was 2.5 minutes.

The Ye-152A prototype with two dummy K-9 missiles on board.

Mikoyan Ye-151

This was the Ye-150 fitted with moving (elevating) cannons. Chapter Two referred to the SN development of the MiG-17 with three elevating 23mm cannon in its nose, an unsuccessful arrangement. For the Ye-151 a special fitting, the DB-66, was designed which involved a revolving ring fitted around the extreme forward portion of the fuselage. This carried two cannon on tilting mountings and was to be capable of rotating through a full 360°, while the gun barrels could tilt + or – 30° as well thus giving a 60° cone of fire. The arrangement gave little drag but did increase the length of the forward fuselage. However, after a full-scale mock-up of the forward fuselage had been built, it was realised that any minor deflections of the cannons could produce forces which would destabilise the Ye-151 along all three axes, thereby creating the potential for a loss of control of the aircraft. Accurate shooting would also be impos-

The Mikoyan Ye-150 prototype had a truly ferocious appearance.

sible and so the Ye-151, and the Ye-151-2 proposal with the 'cannon ring' moved aft of the cockpit, were dropped.

Mikoyan Ye-152A

Problems with the development of the Tumanskiy R15-300, and the introduction of the K-9 AAM into the picture, prompted the Council of Ministers to issue new directives in April and June 1958 that now requested just two Ye-150s, two examples of a new version called the Ye-152 with a single R15-300, and one Ye-152A powered by two smaller Tumanskiy R11F-300 units. The Ye-152s would carry K-9 missiles. The two -152 variants differed only around the rear fuselage and, although their forward fuselage was different to the Ye-150, the wings would be the same. Ye-152A would also carry the 'private venture'

The first Mikoyan Ye-152 prototype.

TsP radar, which had proved superior to the Uragon-5's equipment. Thanks to the delays noted with the first Ye-150 the Ye-152A became airborne first, flying on 10th July 1959, and it showed superb performance coupled with reliable engines. Service ceiling was 19,800m (64,961ft), the aircraft took just 1.48 minutes to get to 10,000m (32,808ft) and top speed with the missile pylons in place was 1,650km/h (1,025mph) at 13,000m (42,651ft). Because it took part in the July 1961 Tushino air display, the only Ye-152A to be built was christened *Flipper* by the ASCC (and initially labelled MiG-23), but in fact this aircraft had been produced purely for research. In 1965, after a four-year career, the Ye-152A was lost in a fatal crash.

Mikoyan Ye-152

The next arrival incorporated changes based on flight experience with the '150' and '152A', such as clipped wingtips, a larger wing area (achieved by reducing the leading edge sweep angle) and missiles mounted at the tips. In addition the development of the R15-300 had hopefully produced a reliable engine – at last! The first Ye-152 flew on 16th March 1961 and showed a 'clean' service ceiling of 22,680m (74,409ft) and took 5.92 minutes to

Model of the Mikoyan Ye-152P/M with canard foreplanes in place.

reach 15,000m (49,213ft) with two missiles, or 4.73 minutes clean; in fact this extraordinary performance capability allowed the aircraft to crack several world speed records. Maximum speed with the missiles aboard was 2,650km/h (1,647mph) at 16,200m (53,150ft). However, the engine still gave problems and some other faults were found in the first aircraft that had to be cured. These brought some changes to the second machine, which flew on 21st September 1961, but the powerplant continued to give lots of trouble. In addition the K-9 missile was suffering its own development problems and some tests also suggested that it would possess an unsatisfactory kill ratio. Consequently the weapon

was abandoned before any trial launches had been made, in a move that also brought an early closure to the flight test programme of the '152'. Plans were made however, to convert the airframe into another new version called the Ye-152P.

Mikoyan Ye-152P and Ye-152M

This advanced variant would receive an improved engine called the R15B-300, complete with a convergent-divergent nozzle that took about 25.3cm (9.96in) off the length of the fuselage, and more fuel in a much deeper

spine along the upper fuselage. There would also be upgraded avionics, including the Uragan-5B-80 radar, to work with the Tu-128's K-80 long-range missile. The airframe's tail was unchanged but the wings had extra tip sections, increasing the span by 1.507m (4ft 11½in), and small mid-span boundary layer fences. Finally, small canard foreplanes were added level with the cockpit to improve longitudinal stability at transonic speeds (these had been tested on a MiG-21F designated Ye-6T-3). The wingtip missile mounts were retained.

The prototype was to be the second Ye-152 converted but the resulting aircraft differed quite a bit from the original proposals and so was renamed Ye-152M; in addition the canards were quickly removed but their fair-

The Mikoyan Ye-152M prototype, complete with huge dummy K-80 missiles on their tip mountings.

Another view of the Ye-152M. Nose the fairing left on the side of the fuselage beneath the cockpit after the canard was removed.

ings were left behind on the fuselage side. Test launches of the K-80 proved to be unsatisfactory because the wingtip rails had insufficient rigidity, which made the weapon 'wobble' after release and lose track of its target. In addition the trials of the Uragan-5B were suspended in 1961/62 because of their complex nature, late development and also the allocation of any further funding to alternative space and missile programmes. Compounding these troubles was the still unreliable engine and so, after the Ye-152M flight test programme was completed, the aircraft was retired.

This finished any plans to put this type of interceptor into service but a tremendous amount had been learnt, not just from flight data but also with experience in manufacturing techniques and materials, much of which found its way into the programme leading to the MiG-25 (Chapter Eight). However, the Ye-152's speed record attempts needed to be registered with the International Federation of Aeronautics and, in a classic Soviet cover-up, the aeroplane was given the false designation Ye-166. The inaccuracy of this bogus title was not confirmed until after the Cold War had ended; the Ye-152M itself went to the Monino Museum.

Sukhoi Rivals

Sukhoi T-37

Sukhoi's competitor to the Ye-150 series was a potentially very impressive and remarkable aeroplane. The development of this new, very fast high-altitude fighter was initiated in early 1958 and it was intended to match the latest Air Force requirement which included a speed of 3,000km/h (1,865mph) at height and a ceiling of 27,000m (88,583ft). Design work commenced after official approval had been given by a Council of Ministers directive issued on 4th June. There were two candidate engines to power the machine, the Lyul'ka AL-11 which was still only a paper design, and the Tumanskiy R15-300 which was already undergoing flight-testing in the Ye-150; the latter being duly selected. Once again the T-37 was to form part of an overall weapon system which here was called T-3A-9 to indicate that it was an improvement over the T-3-51 (Su-9) covered by Chapter Three. The weapons were to be two Mikoyan K-9 semi-active radar homing AAMs on underwing pylons and guided by the TsP-1 fire control radar housed in the intake shock cone, in other words the same Uragan-5 control system as used on the Ye-152.

By the mid-1950s Sukhoi had become firmly stuck on the delta wing and in fact continued to use it on new designs for a long time. It was thus no surprise that this fighter had a delta, in fact it shared pretty well the same configuration as the original T-3 (the wing was swept 60° on the leading edge and had a t/c ratio varying between 4.2% and 4.7%). By employing the same planform the aircraft would also share the proven aerodynamics of the T-3 and T-43 prototypes but the new powerplant, and the high speeds that it offered, required other advances in technology. Once again the inlet was crucial and at the advanced development project stage the fixed centre cone was very long, the inlet itself being axisymmetrical (as built the cone was a little shorter but still of complex design). Instead of employing the centre cone, airflow control was to be secured by a translating outer ring on the forward fuselage together with auxiliary blow-in doors placed in the side of the fuselage.

Much use was made of titanium alloys and steel, particularly in the rear fuselage, and the fuselage structure itself did not use normal stringers but instead just four longerons and monocoque skinning, in a format designed by K A Kuriyanskiy. High-strength aluminium alloys were used elsewhere. Estimated service ceiling in sustained level flight using the afterburner was 25,000m to 27,000m (82,021ft to 88,583ft) and the alternative weapon loads comprised 57mm folding-fin rocket projectiles in underwing pods or 212mm high-velocity rockets on launch rails. Chief project engineer was I E Zaslavskiy.

The T-37's advanced development project was completed in the early spring of 1959, construction of the first of three prototypes began in the summer and by February 1960 the airframe was in its jigs. However, Khrushchev's fascination with missiles has featured several times in these pages and the T-37 proved to be another victim of that policy, though the development problems suffered by the Uragan-5 did not help. In the summer of 1961 Sukhoi was ordered by GKAT, the State Committee for Aviation Equipment, to stop all work on the project, despite the fact that the prototype was pretty near complete. All of the hardware so far produced had to be destroyed and consequently the airframe was taken out of its jigging and scrapped. This must have been a sad decision for Sukhoi but one wonders how the development problems suffered by the Ye-150 series' engines might have affected

Artist's impressions of the Sukhoi T-37 taken from the advanced development project documents.

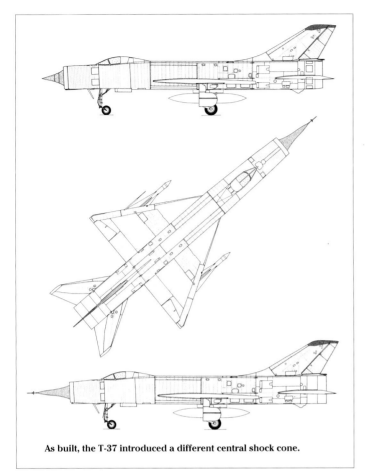

As built, the T-37 introduced a different central shock cone.

Model of the T-37. John Hall

the T-37. By the time the programme was closed it is possible that the weapon system had been retitled T-9M.

Some time later the T-37 managed to exert an influence on the development of another new Sukhoi aircraft, the T-4 bomber, which was also to be built in titanium. In the late 1960s parts of the T-37 prototype were still lying around in the factory and these were inspected in readiness for the T-4's construction. They revealed cracking in some titanium welds due to hydrogen embrittlement. Basically when titanium alloys are molten or very hot, such as during the weld process or the heat treatment operation designed to give the metal its optimum properties, they absorb hydrogen from the atmosphere. Too much gas absorption (literally measured in just hundreds of parts per million) will make the metal brittle. Shock resistance is thus reduced and, consequently, the sharp application of heavy loads can form cracks and allow them to propagate – a very undesirable situation.

As a result automated welding and large vacuum furnaces had to be introduced for the T-4 programme, at considerable cost. One is left wondering therefore what might have happened to the T-37's airframe during an extended flight test programme, had it taken

place. At best there may have been a need to replace parts after a relatively short time in the air, at worst there might have been an in-flight structural failure. In the late 1950s, when work began on the T-37, the fabrication of titanium for aerospace purposes was a very new art and much had still to be learnt; by the late 1960s most of that technology had matured.

Sukhoi P-37

Alongside the T-37 with its nose intake, Sukhoi worked on this alternative layout fitted with lateral intakes. No information is currently available but it would appear that the project was short-lived.

Yakovlev Proposal

Yakovlev Yak-33

To round off this chapter it is worth taking a quick look at this multi-role proposal from Yakovlev. When it was described in the sister volume on bombers much of the material covering it was still classified – a little more has now become available. In the early 1960s the Yakovlev OKB embarked on what was to become a family of aircraft collectively referred to as the Yak-33. Sharing a common

airframe but differing in mission equipment, this family embraced bomber, interceptor and reconnaissance types. The aircraft was configured for high-speed low-altitude operations and would have supersonic performance and VTOL capability, a concept that today might well be regarded as having been ahead of its time. Yak-33 used a series of alternative basic layouts comprising a version with high-set delta wings and a traditional tail unit (of which little is known), a tailless delta-wing type with high-set wings, and another with a delta wing plus a canard.

The tailless-delta had two cruise engines placed side-by-side in the rear fuselage, which are believed to have been Kolesov RD36-41s with a nominal rating of 68.6kN (15,435 lb) dry and up to 156.8kN (35,280 lb) in reheat. They were fed by lateral air intakes and for take-off the jet exhaust would be diverted downwards through nozzles placed ahead of the afterburner, thus giving extra lift. These were supplemented by six 29.4kN (6,615 lb) lift engines placed between the air intake ducts. The canard version had two RD36-41 cruise engines, again with thrust-vectoring nozzles, housed in nacelles attached to the tips of the delta wing. Each nacelle also contained two lift units and, in addition, a pair of lift engines was accommodated side-by-side in the midships fuselage behind the canard foreplanes. Both types had Yakovlev's favoured bicycle undercarriage with outriggers at the wingtips, the latter retracting either into special fairings (on the tailless-delta) or the nacelles (on the canard). All versions had pilot and navigator seated in tandem.

Dimensions were (tailless delta first, canard in brackets) length 26.375m/86ft 6½in (27.0m/88ft 7in) and span 10.25m/33ft 7½in (11.1m/36ft 5in). Both designs had a maximum take-off weight of 40,000kg (88,183 lb), a cruise speed of Mach 2 and a maximum speed of Mach 3. During 1962 the Yak-33 bomber was a contender in a supersonic strike aircraft competition. It carried a weapon load comprising a Kh-45 anti-ship missile or various free-fall stores but was rejected because it failed to meet some of the stipulated performance figures; this decision would most likely have contributed to the lack of interest in the interceptor. However, during a press conference held at the 1967 Paris Air Show Aleksandr Yakovlev declared that his newest fighter was capable of speeds in excess of 3,000km/h (1,865mph), but he did not identify the aircraft in question. He was almost certainly talking about the Yak-33 but the USSR's need for a very fast interceptor was shortly to be filled by the Mikoyan MiG-25 (Chapter Eight).

Heavy Interceptors – Data / Estimated Data

Project	Span m (ft in)	Length m (ft in)	Gross Wing Area m² (ft²)	All-Up-Weight kg (lb)	Engine kN (lb)	Max Speed / Height km/h (mph) / m (ft)	Armament
Lavochkin La-250 (flown)	13.9 (45 7)	26.8 (87 11)	80.0 (860.2)	27,500 (60,626)	2 x VK-9 117.6 (26,455) reheat 2 x AL-7F fitted 98.0 (22,045) reheat	VK-9 at 12,000 (39,370) 1,700 (1,057) clean, 1,600 (994) with missiles	2 x 275 AAMs
Tupolev Tu-128 (flown)	17.53 (57 6)	30.06 (98 7.5)	96.94 (1,042.4)	43,000 (94,797)	2 x AL-7F-2 66.6 (14,990), 99.0 (22,270) reheat	At height: 1,910 (1,187) clean, 1,665 (1,035) with missiles	2 x R-4R + 2 x R-4T AAM
Mikoyan I-3	8.978 (29 5.5)	14.83 (48 8)	30.0 (322.6)	8,954 (19,740)	1 x VK-3 56.1 (12,630), 82.7 (18,600) reheat	1,274 (792) at S/L, 1,775 (1,103) at 10,000 (32,808)	3 x 30mm cannon
Mikoyan I-3P	8.978 (29 5.5)	?	30.0 (322.6)	9,790 (21,583)	1 x VK-3 56.1 (12,630), 82.7 (18,600) reheat	Unknown	2 x 30mm cannon, 16 x 57mm RP or 2 x 190mm RP
Mikoyan I-3U (flown)	8.978 (29 5.5)	15.785 (51 9.5)	30.0 (322.6)	9,220 (20,327)	1 x VK-3 56.1 (12,630), 82.7 (18,600) reheat	1,960 (1,217) at height	2 x 30mm cannon, 16 x 57mm RP
Mikoyan I-7U (flown)	9.976 (32 9)	16.925 (55 6.5)	31.9 (343.0)	11,540 (25,441)	1 x AL-7F 62.9 (14,150), 90.2 (20,300) reheat	1,420 (883) at height	2 x 30mm cannon, 4 x 4 x 57mm RP
Mikoyan I-7K	9.976 (32 9)	?	31.9 (343.0)	?	1 x AL-7F 62.9 (14,150), 90.2 (20,300) reheat	?	4 x 4 x 57mm RP, 2 x K-6 AAM
Mikoyan I-75 (flown)	9.976 (32 9)	18.275 (59 11.5)	31.9 (343.0)	11,470 (25,287)	1 x AL-7F 62.9 (14,155), 90.2 (20,305) reheat	2,050 (1,274) at 11,400 (37,402)	2 x K-8 AAM
Mikoyan I-75F	9.976 (32 9)	18.275 (59 11.5)	31.9 (343.0)	11,380 (25,088)	1 x AL-7F-1 62.9 (14,155), 90.3 (20,315) reheat	2,360 (1,467) at 18,000 (59,055)	2 x K-8 AAM
Mikoyan Ye-150 (flown)	8.488 (27 10)	18.140 (59 6)	34.615 (362.2)	12,435 (27,414)	1 x R15-300 66.6 (14,990), 99.4 (22,375) reheat	2,890 (1,796) at 19,100 (62,664) (Mach 2.65)	2 x K-7S or K-6 AAM
Mikoyan Ye-152A (flown)	8.488 (27 10)	19.00 (62 4)	32.02 (344.3)	13,550 (29,872)	2 x R11F-300 38.0 (8,555), 56.2 (12,655) reheat	2,135 (1,327) at 13,700 (44,948) 2,500 (1,554) at 20,000 (65,617)	2 x K-9 AAM
Mikoyan Ye-152 (flown)	8.793 (28 10)	19.656 (64 6)	40.02 (430.3)	14,350 (31,636)	1 x R15B-300 66.6 (14,990), 100.0 (22,510) reheat	3,030 (1,882) at 15,400 (50,525)	2 x K-9 AAM
Mikoyan Ye-152P/M (flown)	10.300 (33 9.5)	19.656 (64 6)	42.98 (461.2)	?	1 x R15B-300 66.6 (14,990), 100.0 (22,510) reheat	2,681 (1,666)	2 x K-80 AAM
Sukhoi T-37	8.560 (28 1)	19.413 (63 8)	34.0 (365.6)	10,750 (23,699) estimated	1 x R15B-300 66.6 (14,990), 100.0 (22,510) reheat	3,000 (1,864) at 15,000 (49,213) estimated	2 x K-9-51 AAM or 57mm RP pods or 212mm RPs

Chapter Six

Vertical Take-Off

Attempts to provide aircraft with the ability to take-off vertically, or at least with a very short take-off (STOL) performance, have occupied designers around the world pretty well since the start of powered flight. Apart from the helicopter, relatively few military designs have achieved anything like success, and in the jet era by far the most fruitful has been the British Hawker Siddeley Harrier. In the Soviet Union, however, the similar Yak-38 design from Yakovlev went part of the way to matching the Harrier. In fact, apart from isolated studies by Mikoyan and Sukhoi, one associates nearly all of the post-1945 vertical take-off combat aircraft designed in the USSR with the Yakovlev Design Bureau. This is indeed true but other OKBs have had a go and most of their efforts are described here. To begin we have a rather unusual one-off design.

K V Shuleikov

Shuleikov VTOL Fighter Project

This remarkable private VTOL aircraft study is believed to have been made in the either the late 1940s or early 1950s, the date being signified by the engine which, from the drawing, is a centrifugal RD-45/VK-1 type. K V Shuleikov was a teacher at the Zhukovskiy Military Academy and this design was his own idea for a jet fighter. At first glance it appears to be rather like one of the early 'pod and boom' fighters from Lavochkin or Yakovlev, but a closer look reveals vectoring nozzles connected to the engine. There is a nose intake and the power unit is mounted at a downward angle in the forward section of the fuselage ahead of the cockpit, which must have given the pilot pretty poor external vision. The

The Yakovlev Yak-41M seen aboard the carrier *Gorshkov*.

exhaust then splits into four with a forward-rotating nozzle on each side of the lower fuselage and split jetpipes beneath the cockpit. For stability in the hover there is a vertical 'nozzle' or 'fan' in the centre of each wing fed by exhaust gas bled from its respective jet-pipe. The artist's impression shows quite beautifully the operation of the nozzles in the forward VTOL position.

A cannon was fitted in the starboard side of the lower nose outside the inlet lip and an 'old fashioned' tailwheel undercarriage was used, the main wheels being housed beneath the wing roots in the lower fuselage. Span was 7.0m (22ft 11½in) and length 9.4m (30ft 10in). For its time this is a fascinating design

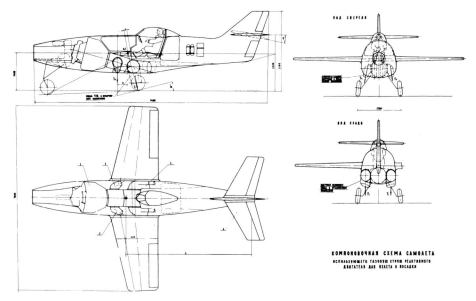

and pre-dates by several years French engineer Michel Wibault's 1956 Gyroptère proposal which also featured vectored-thrust nozzles. Like Shuleikov's design that aircraft was a flat-riser, as opposed to various alternative tail-sitter VTOL types studied during the 1940s and 1950s, but it had four nozzles which were evenly disposed around the CofG. Although not militarily attractive, the Gyroptère was to prove part of the inspiration behind the Harrier – sadly for Shuleikov, his project was stillborn. At one stage Shuleikov also worked for Mikoyan and he rejected this VTOL project as a possible future design.

'Flying Bedstead'

Rafaelyants Turbolyot

During the late 1950s and through the 1960s both France and Britain gained a lot of experience in V/STOL-type aeroplanes. They also gathered some very early experience using test-rig airframes fitted with engines whose thrust was directed downwards to allow the machine to hover above the ground, but which were never intended to fly any distance. The British built the Rolls-Royce Thrust Measuring Rig or 'Flying Bedstead', which flew in mid-1953, while the French built two test machines. There was the EWR Sud hover rig fitted with three Rolls-Royce RB.108 lift jets and then in 1957 came the Snecma ATAR-Volant which served as a testbed for the Coléoptère research machine. The USSR did something very similar.

Aram Nazarovich Rafaelyants was the chief engineer of the Civil Air Fleet (GVF) repair and modification shops at Bykovo and after 1945 was dealing constantly with jet-powered projects and their testing. It was the 'Flying Bedstead' that inspired him to produce an equivalent called the Turbolyot, which had a 63.7kN (14,330 lb) Lyul'ka AL-9G engine mounted vertically in the middle of a frame built of steel tubing. The 'intake' was on the top and fixed to one side of the engine was an enclosed cab for the pilot. There were four main structural girders, each having its own landing leg plus a pipe running along the top with upward- and downward-pointing 'puffer' nozzles on the end. A bleed system from the engine fed each pipe and the nozzles were used to stabilise the aircraft via a modulating system operated on the control column.

Above: **K V Shuleikov.**

Right: **Superb artist's impression of the Shuleikov Fighter in VTOL mode.**

Below: **The Rafaelyants Turbolyot engine testbed.**

Like the initial tests of many VTOL-type research aircraft worldwide, the Turbolyot's first 'flight' trials were made tethered to the ground, or in this case a gantry. Those initial tests were made in early 1957 and by October of that year progress was such that the Turbolyot could be demonstrated publicly in free flight. The Turbolyot was actually a civilian research machine but the data acquired proved very useful for the Yak-36 military project described shortly. Today the Turbolyot still survives in the Monino Museum.

Yakovlev VTOL Series

Of all of the Soviet Design Bureaux it was the Yakovlev OKB that really got its teeth into VTOL and STOL combat aircraft design. This team was to produce a long series of projects stretching from early basic research aeroplanes through to supersonic fighters and all known projects resulting from this work are described below. Some coverage was given to the Yak-36/Yak-38 series in the sister volume on bombers because the latter was eventually developed as an attack aircraft. Consequently the authors must apologise for any repetition here, but when work began on these types the primary objective was to create fighter-type aeroplanes.

Yakovlev Yak-V and Yak-36

In September 1960 Aleksandr Yakovlev visited the Farnborough Air Show where he was

Above: **Yakovlev Yak-36 second flight article.**

Left and below left: **This proposed development of the Yak-36 had a small nose radome.**

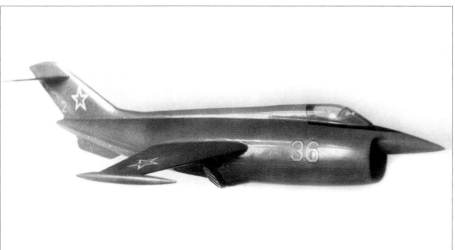

impressed and inspired by the performance of the British Short SC.1 VTOL research aircraft. This aeroplane had fixed vertically-mounted engines for lift and another conventional unit to supply forward thrust. On his return to Moscow Yakovlev made recommendations to the Soviet deputy prime minister responsible for defence, Dmitri Ustinov, that a Soviet VTOL research programme should be started. In fact, with America, Britain and France all by now seriously considering VTOL combat aircraft, getting Soviet Air Force officials to follow suit was not too difficult. By the end of year the Yakovlev design team had begun studies for a V/STOL research aeroplane and the initial examinations were based on the Yak-30 two-seat jet trainer. The Yak-30V (*vertikalnyi* or vertical) had two vertically-mounted lift engines to go with its normal propulsion unit – all three of them Tumanskiy RU-19s giving 10.3kN (2,315 lb) of thrust.

However, the winter of 1960 and spring of 1961 saw the first hover and flight trials of the Hawker P.1127 research aircraft (which eventually led to the Harrier) and this used rotating vectored-thrust nozzles to generate both downward thrust for hovering and propulsive thrust as required. On seeing this Yakovlev discarded the Yak-30V and switched to a new design with vectoring nozzles. On 30th October 1961 a Council of Ministers directive was issued ordering the OKB to design and build a

Soviet Secret Projects: Fighters Since 1945

Yakovlev Yak-38.

single-seat 'fighter-bomber'. This would be powered by two 49.0kN (11,025 lb) R21M-300 engines and was to have a maximum speed of around 1,200km/h (746mph), but in due course it was established that a better engine was required and the choice settled on the unreheated R27V-300 which was designed by Tumanskiy and Khatchaturov. This powerplant was chosen because, at this time, there was no suitable engine available that could supply sufficient thrust to operate four vectored nozzles.

The aircraft project was originally called the Yak-V, before being renumbered Yak-36. Its design takes us back to the early 1950s and the general configuration of the MiG-15 and its sisters, but this aeroplane was a lot heavier. There was just the one rotating nozzle on each side while puffer pipes for control and stability were directed to the wingtips, the nose (in a long boom extending well over the top of the intake) and below the fin. A long boom had to be used because, to keep down weight, the fuselage itself was quite short.

Four Yak-36 research aircraft were built, although this number includes the static test airframe. The first flight specimen was completed in 1962 and its first hovering tests were made, tethered to the ground, on 9th January 1963. A full transition to horizontal flight from a vertical take-off, and then back to the hover for a vertical landing, was completed on 24th March 1966. These prototypes gave a lot of experience in VTOL design but on their own they displayed a poor performance; for example their ceiling was only around 12,000m (39,370ft). Some brief trials were made with a Yak-36 aboard the helicopter cruiser *Moskva* and an example took part in the July 1967 Domodedovo air display. This revealed the type's existence to the public and soon afterwards the Yak-36 was given the Western codename *Freehand.*

Yakovlev Yak-36M and Yak-38

The Yak-36's performance contributed to an Air Force decision not to fund further VTOL studies, which meant that an offer made by Yakovlev to build a trial batch of aeroplanes was not accepted. However, the Soviet Navy was planning a new design of aircraft-carrying cruiser called Project 1143 (the *Kiev* class)

and was interested in the potential benefits of having a VTOL type on these ships. Something better than the Yak-36 would be needed, however, and so the OKB continued its studies and by the mid-1960s it was looking at a much-improved project called the Yak-36M. The initial layout actually looked like a slimmed-down Yak-36 and had side intakes level with the cockpit and two RD-27VM-300 jets. However, after a great deal of research into lift jets and the techniques that were required to control the powerplant as a whole (which saw some alternative arrangements rejected), the final design had just a single main cruise engine, two rotating nozzles and two vertical lift jets. The latter were the new RD36-35 units designed by the Kolesov Bureau. The Yak-36 had also used Yakovlev's favoured bicycle undercarriage but on the -36M this had disappeared to be replaced by a tricycle format.

There were still many officials who doubted the wisdom of having a VTOL aircraft but it was the Minister for Aircraft Industry, Pyotr Dementiev, who kept things going with a suggestion that the Yak-36 should be used as the start point for a two-stage development programme. The first portion would be the Yak-36M light attack aircraft fitted with basic avionics, and then this would be followed by a much more sophisticated supersonic fighter called the Yak-36MF or Yak-36P. Incidentally, the same Yak-36P designation was also allocated to a pure air defence interceptor variant of the Yak-36M with three RD36-35 engines. At the end of December 1967 a Council of Ministers directive ordered that five Yak-36M prototypes should be built and then on 25th January 1968 the Air Force's Commander-in-Chief also approved a requirement for the light attack version of the Yak-36M. The Navy had become the driving

A proposed supersonic derivative of the Yak-38, which was actually an attack aircraft design, that represents one of the earliest steps taken towards producing the Yak-41. It has box air intakes and a single afterburner nozzle.

Yakovlev Yak-39

In service the Yak-38 experienced a lot of problems that required many improvements. One effort intended to enhance the basic aeroplane was the Yak-39 V/STOL fighter/bomber designed in 1983 by Mr Bekerbayev. Despite having a similar appearance to the Yak-38 overall, this introduced a new larger and thicker wing built in composite materials, a different tail and the capability to carry more weapons and internal fuel. The wing would not produce any improvement in performance but it did offer more space for fuel plus extra lift. Yak-39 had a single 65.6kN (14,770 lb) R28V-300 lift/cruise engine with swivelling nozzles, two 40.2kN (9,040 lb) RD-48 lift jets and an S-41D radar which needed a large radome to cover the scanner. Cruising weight was 12,550kg (27,668 lb), a 30mm cannon was installed and the estimated performance figures included a maximum 900km/h (559mph) at low level and a range of 450km (280 miles). In July 1983 the Yak-39 was rejected by an official commission because it was realised that the type's limited air-to-air combat capability would prevent it from attacking more than a single subsonic air target at any one time. This decision was also influenced by the existence of the Yak-41 project offering supersonic performance and the fact that the Yak-39 still showed many of the Yak-38M's weaknesses. Work on the Yak-39 was terminated immediately.

Yakovlev 'Yak-41' Pre-Projects

The supersonic Yak-36P noted earlier was never built, but the next stage in this story was to be the development of a supersonic V/STOL fighter. A contract to produce such an aircraft was placed in 1975 and, not surprisingly, a series of alternative designs were studied under the in-house designation 'Article 48'. These featured several combinations of lift and cruise engines and the project as a whole was officially called Yak-41, with the intention that it should join the Soviet Navy. One of the earliest proposals was based on the Yak-36 and had a large fat fuselage for the engine, a relatively high wing position, a small nose radome and an R-27 missile under each wing. The next layout to be assessed was a development of the Yak-38 with a single lift/cruise engine, a chin intake geared for supersonic flight with a semi-circular centre cone and a single vectoring nozzle. It was a single-seater and had a small cranked delta

force behind the project but the Air Force still retained overall responsibility.

A mock-up was examined officially on 15th April 1970 and the first of what became four prototypes made its first free hover on 22nd September 1970, and then a maiden conventional flight on 2nd December. On 25th February 1972 a full-profile flight was completed for the first time and on 18th November one of the prototypes made the first landing on the deck of a warship – again the *Moskva* was the trials ship. However, thanks to some development problems, it was not until October 1976 that the aircraft was introduced into service (with the designation Yak-38). Although the Yak-38 became primarily an attack aircraft, it did carry two Bisnovat R-60 AAMs for Fleet defence duties. Other stores included a

gun pack, rocket projectiles, bombs of up to 500kg (1,102 lb) weight and even a 250kg (551 lb) free-fall nuclear weapon. The upgraded Yak-38M was first flown in November 1982 and could carry more weapons and had a more powerful engine. In all 193 single-seat Yak-38s were built together with 38 two-seaters. The last flights were made in 1991 and in the West the aircraft was known as *Forger*.

Yakovlev Yak-36-70F

This was a study undertaken in 1970 for a light supersonic VTOL fighter that featured two afterburning lift/cruise engines with variable inlets and no lift engines. It also had a bicycle undercarriage, but the project stayed on the drawing board.

wing, combined mid-wing nacelles/pylons and a rear fuselage that was quite similar to the Yak-38's.

Another 'competitor' was a V/STOL development from the Yak-45 project that in 1972 had competed against the designs that led to the Mikoyan MiG-29 and Sukhoi Su-27 (Chapter Nine). This version had two RD-38 lift jets in the fuselage, twin 'Article 69' main engines in wing nacelles and rectangular vectoring nozzles on the end of each nacelle. It also had a high cranked delta wing, an even higher-placed trapezoid canard behind the cockpit canopy that almost overlapped the start of the wing leading edge, a narrow-track tricycle undercarriage and wingtip-mounted AAMs. The project also posed the question of how to survive the failure of any engine when oper-

ating in the hover mode; this was never satisfactorily answered and so the Yak-45 development was dropped.

All of the remaining studies had twin fins whereas those described so far had just a single fin. The first had a long nose to accom-

modate lift engines in front of the cockpit, side intakes for the main engine, a wing swept on the leading edge but with a near-straight trailing edge, leading edge root extensions, low set tailplanes and a tricycle undercarriage. A model also shows it carrying a single Bisnovat R-27 AAM under each inner wing. The next project was the first to suggest the Yak-41's final configuration but this had semi-circular side intakes set high on the fuselage, outwards canted fins set at a much higher angle than on the Yak-41, and a tricycle undercarriage. In addition, instead of lift engines this design had a large protruding rectangular afterburning 'chamber' which was to be directed downwards for VTOL operation.

A later unbuilt project came very close to the Yak-41 itself and was modelled in two forms. The first had semi-circular side intakes rather than the wedge-shaped box format used on the prototypes while the second was near-identical except that it did have the box intakes. Both models had an R-27 on each inner wing pylon and a smaller AAM on each outer (mid-wing) pylon. All of these later projects could fold their wings at around the mid-wing position but the Yak-45 development could only fold the portion available outside its nacelles. Eventually this substantial research into V/STOL fighter configurations was to lead to some hardware.

Yakovlev Yak-41

As built the Yak-41 had two Rybinsk RD-41 lift engines fixed in tandem just behind the cockpit at an inclined angle of 85° and for the main engine there was a new two-shaft turbofan called the R-79V-300. This was designed for the aircraft by the Soyuz Design Bureau (formerly the Tumanksiy Bureau, the name was changed after Tumanksiy died in 1973). The

Model of the Yak-45 development (1978/79).

Left and below: **Late Yakovlev Yak-41 model, which comes very close to the Yak-41 as built. There were two versions of this model, one with semi-circular intakes and the other (as shown) with wedge intakes.**

for the first time on 26th September 1991. Some impressive performance figures were achieved, including a top speed of Mach 1.74 and a maximum rate of climb of 15,000m (49,213ft) per minute. These allowed the Yak-41M prototypes to break several time-to-height records (under the fictitious designation Yak-141). The type's service ceiling was 15,000m (49,000m).

The end of the Cold War, however, brought an end to the programme because in November 1991 all funding for it was suspended by the government. As a result the Yakovlev team had to look for alternative financing and in September 1992 it brought one of the prototypes to the Farnborough Air Show in England. No weapons were displayed at the show and it appears that few, if any, were ever carried, although the last prototype was photographed in Russia with two R-27 AAMs on the inner underwing pylons and two Vympel R-73 short-range AAMs on the outer pylons. At Farnborough the aircraft generated a lot of interest but no overseas money was ever forthcoming and soon afterwards the project was abandoned. In the autumn of 1994 there were reports that it would be resurrected. One prototype had been damaged in a landing accident and had never been repaired, but the other was apparently still airworthy; however, nothing more appears to have been done. The damaged airframe has since been scrapped but the other has been preserved.

The Yak-41 remains one of very few V/STOL combat aircraft designs to have flown supersonically on the level. It was a complex machine that may have given problems in service, but from a technical point of view it was also very interesting and it is a pity that no-one knows how well it would have performed. Both the Yak-41 and the Yak-38 were very noisy aeroplanes and the former needed reheat to accomplish a vertical take-off. The supersonic aircraft was to become an early victim of the Russian budget cuts introduced in the wake of the Cold War that gave the country great problems in the development and procurement of new defence equipment. As we shall see in Chapter Nine, there would be other cancellations and today getting sufficient finance for armaments is still a problem. In more recent times Yakovlev's V/STOL knowledge has been employed on the new American Joint Strike Fighter project.

R-79V-300 had a rotating tail jetpipe, which could be set 95° down for a vertical take-off and at 63° for a short take-off, and augmented thrust which gave ratings between 107.8kN (24,250lb) dry and 151.9kN (34,170lb) with maximum augmentation. The maximum available thrust in VTO mode from all of the engines was 217.5kN (48,940lb), which included 137.2kN (30,865lb) from the main unit.

A decision to go ahead with the Yak-41 was made in 1977 with state testing expected in 1982. Progress, however, was slow and this date was twice postponed. Then by 1985 the design had reached a stage where the aircraft had become much more complex and it was considered that the cost of producing flight articles could be justified more easily if the aircraft had multi-role capability. As a result a new designation, Yak-41M, was issued for a version fitted with considerably more advanced avionics, including a large radar, and the capability to carry an extensive array

of weapons including missiles, bombs and rocket pods. Four underwing pylons were available to carry up to 2,000kg (4,409lb) of stores and a 30mm cannon was to be fitted under the port intake box. In service the Yak-41M was expected to be a multi-role type operating as an interceptor and as an attack aircraft against ground and naval targets. It was decided to build this more advanced version, with a first flight required now in 1988, and drop the original Yak-41 label. The new aircraft carrier *Admiral Gorshkov* was expected to take some pre-production aeroplanes to sea for the first time in 1991/92.

Four airframes were ordered, including a powerplant testbed and static test airframe, and the first prototype made its maiden flight (as a conventional aircraft) on 9th March 1987. Hovering trials (with the second machine) began on 9th December 1989 and a full transformation from hover to forward flight was accomplished on 13th June 1990; both prototypes landed on the *Gorshkov*

Right and below: **Another Yak-41 project model showing a design which, in appearance, is very close to the prototypes.**

Bottom: **Prototype Yakovlev Yak-41M in the hover.**

The Yak-41 also received a Western code-name, *Freestyle*, and attempts were made to improve the aircraft using knowledge gained from flight experience. These included a wider integral fuselage of new shape combined with a delta wing and much larger LERX, but with few changes to the powerplant except in the area of the intakes; additional fuel would have been carried in the new fuselage. This version was really intended for land-based operations with rather more emphasis placed towards short take-off and vertical landing (STOVL) performance.

Yakovlev Yak-43

In 1986 a proposal was made for a new development fitted with a variant of the Kutznyetsov NK-321 turbofan, which was the engine used to power the huge Tupolev Tu-160 inter-continental bomber. This project was called the Yak-43 and the initial work had begun in 1983/84 as a follow-on to the Yak-41M. It used the NK-321 as a lift/cruise engine complete with a vectoring nozzle but, mounted forward, there were separate vertical lift jets. However, the lift units were supplemented in the nose by an auxiliary combustion chamber that received bleed air from the main engine; in reheat the NK-321 would supply a total of 244.8kN (55,070 lb) of thrust. Yak-43 had a wing of trapezoid shape mounted high on the fuselage and with a cranked leading edge, there was no tailplane and twin fins were fitted. The data is believed to include a maximum VTO take-off weight of 15,800kg (34,832 lb) and STO weight of 21,500kg (47,399 lb), the maximum VTO weapon load was 1,000kg (2,205 lb) and STO load 4,200kg (9,259 lb) and the internal fuel totalled 6,000kg (13,228 lb); combat radius after a short take-off was estimated to be about 900km (559 miles).

This new development promised quite an improvement over the Yak-41M in terms of performance and overall capability. Bigger wings ensured more agility, a greater weapon load could be carried over a greater range and the type was expected to perform as a supersonic multi-role aircraft from both land and ship bases. The Yak-43 was also intended to incorporate stealth technology and more advanced avionics but the project stayed on the drawing board, in part because the Yak-41 itself had been shelved.

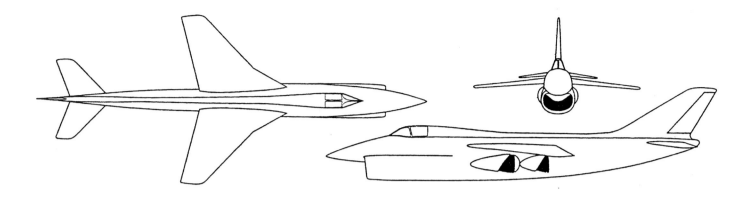

was a large nose intake, with a large radome above, and a pair of closely-spaced vectored-thrust rotating nozzles on each side of the centre fuselage. In appearance the '136' had much in common with the Hawker P.1127 (although that type had side intakes) and was a single-seater; Yakovlev's Yak-V project was also similar to the '136'.

'Aircraft 136' had a span of 5.12m (16ft 9½in) and was 11.8m (38ft 8½in) long. It was to be capable of subsonic speeds only and, to date, no information is available regarding the powerplant. The preliminary proposal and a model of the fighter were submitted to the Air Force for consideration, but the project progressed no further. There is no evidence to suggest that 'Aircraft 136' was in direct competition with any of Yakovlev's early VTOL projects, but it fits in with their timescale very well. In the mid-1970s Tupolev re-used the '136' designation for a passenger aircraft that was intended to operate on trunk routes.

Mikoyan and Sukhoi V/STOL Studies

Mikoyan 23-31 (or Ye-7PD)

The first attempt by the Mikoyan OKB to create a V/STOL-capable aircraft used fixed lift jets rather than vectoring nozzles. This was a MiG-21PFM airframe fitted with two Kolyesov single-shaft RD-36-35 lift jets amidships and a Metskhvarisvili R-11F2-330 flight engine, and it was called the Ye-7PD (PD for *podyomnye dvigatyeli* or lift engines). The idea of using fixed lift jets was studied extensively during the 1950s and 1960s by many aircraft designers worldwide, but particularly so in Europe. It was hoped that lift jets, designed to give short term lift, would be lighter in weight than normal units and some estimates suggested

Tupolev V/STOL Studies

Tupolev 'Aircraft 136'

Although the Tupolev Design Bureau's normal workload concentrated on large heavy aircraft like bombers, airliners and transports, just occasionally it had a go at designing 'non-standard' types like missiles and this front-line multi-role VTOL fighter project. It was

produced in response to an idea from, and against orders raised by, Andrei Tupolev himself and the OKB undertook the study during 1963 and 1964. It was called 'Aircraft 136' and a preliminary technical proposal was quickly prepared. This described an aeroplane with a small high-mounted 'cranked triangular' low aspect ratio wing which was swept 60° on the inner leading edge and 45° further out. There

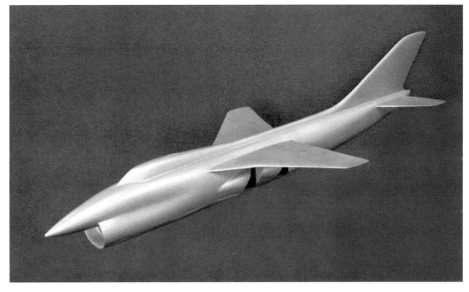

that they would be capable of lifting up to twenty times their own weight. In addition, in the case of Ye-7PD and some other designs, they did not point straight down for a vertical take-off but instead were angled slightly to reduce the take-off and landing distance. After take-off the lift units would be switched off until it was time to land.

The 23-31's lift units were mounted just ahead of the aircraft's CofG inside a new fatter centre fuselage and were angled at 80°. The MiG-21PFM's forward and rear fuselage was retained and the undercarriage was fixed. The lift engine air came through a louvred panel in the upper fuselage aft of the cockpit that closed when the lift units were inactive, and the exhaust exited under the fuselage through a vectoring box which contained seven pivoted transverse deflectors for each engine; this would allow the pilot to deflect the jet flow through a limited angle. No design effort was put into streamlining the airframe or reducing fuselage drag.

This one-off aircraft made its first flight on 16th June 1966 and was demonstrated to the public at the Domodedovo air show in July 1967. Flight-testing showed that the 23-31 was basically unstable at low airspeeds, its control was inadequate and achieving a good landing

Mikoyan 23-31 prototype seen in the hover and displaying the open louvred door. Note also the broader centre fuselage.

was hard work. After Domodedovo the machine was quickly grounded. The confusion created by the type's different designations, which also included 'Izdeliye 92' as used in official circles, was compounded when the western press dubbed the aircraft MiG-21PD (and *Fishbed-G*). The Soviet authorities deliberately chose not to correct the error, which thus implied to the rest of the world that the type was intended for service use. On retirement the airframe was donated to the Moscow Aviation Institute but has since disappeared. The motivation behind producing this aircraft was to prove a concept and gain manufacturing and flight experience with a V/STOL type, rather than any serious attempt

to build a fighter prototype. Mikoyan built another VTOL aircraft called the 23-01, but this is so closely involved with the story of the MiG-23 that it is described in Chapter Eight.

Sukhoi T-58VD

Sukhoi's answer to producing a STOL-technology demonstration aircraft was also to adapt a production type, in this case the Su-15 (Chapter Four). One reason for the OKB undertaking such research was that Sukhoi was considering lift engines for a new strike aircraft design which needed to possess a short field performance. This project became the fixed-wing T-6 prototype fitted with four RD-36-35 lift units but it was later superseded

by a variable-geometry version which became the Su-24. An Su-15-based demonstrator was launched by a MAP order of 6th May 1965 which requested the construction and testing of a proof-of-concept aircraft. Documents for converting a T-58 were soon prepared and these showed three RD-36-35 engines inclined 80° in a new centre fuselage section. There were two scoop intakes on the upper fuselage covered by louvre doors and some other changes, including to the wings, although as a whole the airframe was little different to a standard aircraft. The version was known as the T-58VD (VD – *vertikahl'nyye dvigateli* or vertical engines).

The airframe selected for conversion was the T-58D-1: that is, the first Su-15 prototype. Work began in January 1965 and the alterations were completed towards the end of the year. Trial engine runs began in December and the machine undertook its preliminary tests tethered to the ground. The first flight was made on 6th June 1966, ten days before the Mikoyan 23-31, and flight experience showed that the extra lift supplied by the Kolyesov units reduced both the speed and the distance required for take-off and landing. However, their position inside the airframe was not ideal and on the approach, at speeds below 320km/h (199mph), problems were experienced with pitch-up; these were solved by switching off the forward lift engine when preparing to land. No data has been published regarding maximum speeds, and the like, but with the same main engines aboard and no major change to the basic aerodynamics, one assumes that they would have been fairly similar to a standard Su-15.

An appearance at the Domodedovo show brought a Western codename (*Flagon-B*) and the test data that was acquired proved valuable in the design of the first T-6 prototype. Field performance was good, but at a price because the aircraft's range was much

reduced – with the lift engines in place there was less space for fuel while more of what was available was used for take-off and landing. Longitudinal stability had also deteriorated and in due course the T-6's layout was eventually redesigned with variable geometry wings. After the T-58VD's flight testing was finished, a couple of weeks after Domodedovo, the airframe went to the Moscow Aviation Technology Institute where it served for many years as a training aid. However, when the Sukhoi Su-27 (Chapter Nine) made its appearance, the need to keep the old T-58VD ended and so the airframe was scrapped.

Sukhoi Tail-Sitter

Sukhoi Shkval

All of the V/STOL aircraft and projects studied so far have been flat risers: that is, while taking off vertically they retained the horizontal attitude associated with normal flight. An alternative method was to sit the aircraft on its tail – a tail-sitter – and take off in a vertical attitude in the same way as a ground-based missile. The real problem here was landing again because the pilot would have to look directly behind him to see the ground. The country that explored this type of aeroplane the most was the United States. In 1954 two competing turboprop-powered prototypes of a convey escort fighter, the Convair XFY-1 and Lockheed XFV-1, made their maiden flights. They were intended to take off from small platforms on merchant ships and consequently both were tail-sitters, but the programme was abandoned in 1956. In late 1955, however, trials began of the Ryan X-13 delta-wing research aircraft powered by a Rolls-Royce Avon axial jet. This had a successful career but the tail-sitting format was never adopted for a service aircraft.

In 1960, some years after the American tests had closed down, the Sukhoi Design Bureau

established within its organisation a separate 'private venture' design team to examine aircraft which had an 'unorthodox and less conventional' layout; it was led by R G Martirosov and was principally made up with relatively young engineers who had recently graduated from the Moscow Aviation Institute. The team's work was favourably received, such that within a short time it had been upgraded into a part-time Design Bureau operating outside normal working hours. One of the most interesting ideas to come from this team was the Shkval VTOL tail-sitter fighter, which dealt with the problem of the view back towards the ground by having the pilot's seat pivoting on its axis. This would simplify the job of controlling the aircraft, reduce the chances of a loss of orientation and, in general, ease the pilot's physical workload.

The Shkval had a sleek slim fuselage with two Tumanskiy engines placed side-by-side, 'conventional' box side intakes and canard foreplanes, and four cruciform mainplanes placed diagonally at the corners of the fuselage. Fitted to each wingtip was a cylindrical tube or 'nacelle' on which the aircraft rested on the ground; these also housed pads and shock absorbers at the bottom/rear end with the rest of the tube being filled with fuel. To help stabilise it during take-off and landing the aircraft used vectored thrust control nozzles that used air supplied by an autonomous compressor, although a back-up system was available using air bled from the engine compressor. It appears that two versions were considered, the principal difference being the position of the canards – one had them on the sides of the intake boxes, the second on the nose. The former, Shkval-1, had a span of about 5.8m (19ft) and length 15.0m (49ft). Performance data is unavailable but, judging from its appearance, the Shkval must have been capable of supersonic speeds. A cannon was housed in the bottom left-side fuselage level with the cockpit.

Following the completion of preliminary calculations, some wind tunnel tests were made using TsAGI's facilities and these gave good results. Then the project was forwarded to Pavel Sukhoi for assessment before attempts were made to see if any official funding would be available to allow it to progress further. To this end the OKB was visited by the first secretary of the VLKSM, Sergey Pavlov with G V Novozhilov from the Ilyushin OKB acting as his technical assistant, and they gave the Shkval some very high

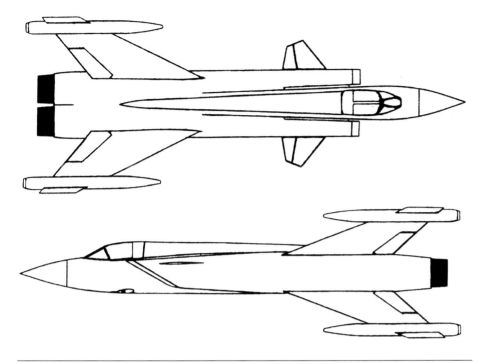

Left: **Sukhoi Shkval-1 (1963).**
Russian Aviation Research Trust

Above: **Models of two versions of the Sukhoi Shkval-1. The model on the right has the canards on the nose (except that one has broken off) and also slightly shorter wingtip nacelles.** Victor Drushlyatov

marks. As a result the VLKSM central committee requested that MAP should allocate sufficient funding to allow the designers to put together a preliminary project, get the opinions of all of the main research establishments (TsAGI, TsIAM, LII and the Zhukovskiy Academy) and also build a full-size mock-up of the forward fuselage.

The funding was made available and the work was completed in six months. The mock-up, which could be pivoted to provide a simulation of the pilot's view in both horizontal and vertical attitudes, was appraised by several pilots and one, Georgi Beregovoy

who would later become a cosmonaut, thought that the operation of an aircraft like the Shkval would be possible. The final report was submitted to MAP for further review and in August 1963 the design team argued its case to the technical experts within the State Committee for Aviation Equipment (NTS GKAT). This meeting was very intensive and the case for continuing the project was strong. However, despite recognising that the Shkval concept had considerable technical merit, the Committee was not in a position to give it the go-ahead, primarily because Khrushchev's and the Soviet Government's

feelings towards aviation over the previous years still exerted their influence. Therefore the Shkval was rejected, but the members of its design team received much praise for their efforts. Indeed, the experience and some of the new elements of design that had been created to bring this fighter design into existence (in several cases new inventions) were later applied to other designs.

V/STOL Aircraft – Data / Estimated Data

Project	Span m (ft in)	Length m (ft in)	Gross Wing Area m² (ft²)	All-Up-Weight kg (lb)	Engine kN (lb)	Max Speed / Height km/h (mph) / m (ft)	Armament
Yakovlev Yak-36 (flown)	7.4 (24 3)	16.75 (54 11.5)	c15.8 (169.9)	9,400 (20,723)	2 x R-27V-300 jet 62.3 (14,020)	1,009 (627) at S/L, 1,000 (622) at 10,000 (32,808)	None but initially intended to carry up to 600kg (1,323 lb) of stores
Yakovlev Yak-38 (flown)	7.022 (23 0)	16.37 (53 8.5)	18.7 (201.1)	rolling take-off 11,300 (24,912), VTO 10,300 (22,707)	1 x R-27V-300 jet 64.7 (14,550) + 2 x RD-36-35 lift jet 28.4 (6,400)	1,150 (715) at 200 (656), 1,020 (715) at 10,000 (32,808)	Max 3,748 lb (1,700kg) stores
Yakovlev Yak-41 (flown)	10.105 (33 2)	18.36 (60 3)	31.7 (340.9)	STO 19,500 (42,989), VTO 15,800 (34,832)	1 x R-79V-300 151.9 (34,170) + 2 x RD-41 80.4 (18,080)	1,250 (777) at S/L, 1,800 (1,119) at 11,000 (36,089)	Intended 1 x 30mm cannon + up to 2,000kg (4,409 lb) stores, including 4 AAM
Mikoyan 32-31 (flown)	7.765 (25 6)	14.72 (48 3.5)	26.5 (285.0)	unknown	1 x R-13F-330 63.6 (14,300) + 2 x RD-36-35 23.0 (5,180)	c600 (373)	None fitted

Chapter Seven

Rocket Power
and Flying Boats

The very impressive-looking OKB-2 '468' rocket fighter. Helmut Walther

ROCKET FIGHTERS

The jet engine brought a revolution to the design of all military aircraft and, from the fighter point of view, the advances in speed made available by the jet put the piston engine out of business. However, until the 1950s jet engines could not supply sufficient thrust to give interceptors a high enough rate of climb to ensure that they could deal thoroughly with oncoming enemy bombers flying at very high altitudes, so for a period of time some alternative sources of power were examined. One potential solution tried by several countries was rocket power and the Soviet Union was no exception. In fact the first reliable rocket engine to be produced in the USSR, designed by Leonid Dushkin, was first tested on the ground in 1939. On 28th February 1940 it was used to power the

Korolyov RP-318 research aircraft in flight, the first time that a Soviet aeroplane was flown under rocket power.

Early Studies

Several rocket-powered fighter and research aircraft projects were under way well before the end of the Great Patriotic War and, for completeness, a brief description of the more important items has been included here.

Bereznyak-Isayev BI

In July 1941 two Soviet designers, Aleksandr Bereznyak and Aleksei Isayev (who both worked at the Bolkhovitinov Design Bureau), produced a proposal for a high-speed rocket-powered interceptor fighter called the BI. The document showed a simple and very small

aeroplane that was built of wood. The official response was an almost immediate go-ahead for five prototypes and on 15th May 1942 the first of these achieved the world's first flight by a fully engineered rocket-powered interceptor. However, problems with the rocket motors and their fuels (red fuming nitric acid plus kerosene) prevented anything more than gliding or towed flight until February 1943, by which time seven prototypes had been authorised. Then on 27th March, after rocket-powered flight had resumed, the third machine dived into the ground when flying at sustained full power. It was found that at a speed of 900km/h (559mph) the BI would pitch nose-down and this weakness helped

Bereznyak-Isayev BI.

The sixth Bereznyak-Isayev BI prototype with a DM-4 ramjet fitted to each wingtip.

to bring an end to plans to build fifty production aeroplanes fitted with a nose gun.

The programme had progressed under conditions of great urgency but by 1944 the need to build production examples of the BI had evaporated. Some of the prototypes, however, were employed right through to the end of the war on various trials, the sixth airframe for example having a Merkulov DM-4 ramjet fitted to each wingtip. The BI was a remarkable achievement but it did not possess anything like sufficient flight endurance to make it a viable project. In 1943 it was estimated that the BI could achieve 1,020km/h (634mph) at height, but after the in-flight crash no attempt was made to find out if this was accurate. Three years after the end of the war Bereznyak proposed a supersonic interceptor powered by an 18.6kN (4,190lb) Mikulin AM-5 turbojet and a 98.0kN (22,045lb) rocket motor, the latter to give a high-speed 'dash' performance of up to Mach 1.8. The estimated range was 750km (466 miles) but the idea stayed on paper.

Florov 4302

Ilya Florov was an engineer who worked with various State establishments and Design Bureaux and during the war he was put in

control of a special design cell that formed part of the Air Force Research Institute (NII-VVS). In 1943 this unit was ordered to design and produce a small rocket-powered research aircraft for testing wing shapes and control systems. The result was called the 4302 and three prototypes were ordered. The first airframe served purely as an unpowered glider for handling tests, for which it was towed to about 5,000m (16,404ft) altitude by a Tupolev Tu-2. The second used a rocket

designed by Isayev and Dushkin which gave 10.7kN (2,425lb) of thrust at sea level, and this aircraft made its first flight in August 1947 and achieved a maximum speed of 826km/h (513mph). However, during August the programme was cancelled and its funding was re-allocated to Mikoyan's I-270 project (below). The 4302s were built in light alloy and Nos 2 and 3 were intended to use a jettisonable take-off trolley and land on a skid. The third airframe was completed and was to

over a year and a half progress was very slow until Stalin livened things up in late 1942 with an official go-ahead. Kostikov was made the project's chief designer (which is why some documents refer to the aircraft as the Ko-3) and two prototypes were approved for manufacture by OKB-55. This Design Bureau was led by Matus R Bisnovat who later became responsible for many of the Soviet Union's air-to-air missiles. Tikhonravov worked on the aerodynamics.

By the spring of 1943 both airframes, built predominantly in wood, were nearly finished but then it was decided to leave off the ramjets and have just the rocket motor for power. The ramjet nacelles were removed, the span was reduced and the aircraft was redesignated 302P (P for *perekhvatchik* or interceptor). Gliding test flights began in August 1943 using a tug to take the 302P to altitude, and these proved that the interceptor flew well. However, the programme was terminated in March 1944 before any powered flying could commence and Kostikov was declared responsible for failing to keep the project on schedule – he was sacked from his position and was later sent to prison. In fact he was an ambitious man who appears to have used Tikhonravov's proposals to gain promotion and power. The 302P was to have had two 20mm cannon fitted in its nose. During the glider trials that were completed it only proved possible to assess the type's handling and collect performance data at low speeds.

be fitted with a more powerful Dushkin and Glushko RD-2M-3 engine; as such it was redesignated 4303 but it never actually received its powerplant.

Kostikov 302 (Ko-3)

A G Kostikov was the Director of the 'Reaction Engine Scientific Research Institute' (the RNII). In 1940 one of his staff, Mikhail Tikhon-ravov, proposed a design for a relatively cheap and simple fighter powered by two ramjets mounted under the wings and a liquid-propellant rocket in the rear of the fuselage. It would lack range but should offer a superb performance, although this theory was to generate a good deal of argument. A preliminary project was approved by RNII's technical experts in the spring of 1941 but for

Polikarpov Malyutka

Before the Great Patriotic War the famous designer Nikolay Polikarpov built a reputation for building fast and agile combat aircraft. This, his last design, was started in June 1943 and was intended to be a short-range interceptor for defending 'high value targets'. The Malyutka or 'Little One' was to be powered by a 9.8kN (2,205 lb) thrust rocket motor designed by the NII Research Institute. It had a single thrust chamber and was fuelled by red fuming concentrated nitric acid and kerosene, which appear to have been the favoured fuels for most of the USSR's rocket engines. To begin with, just the one prototype was planned and Polikarpov monitored its progress directly, construction beginning early in 1944 using a wing and tail made with stressed alloy skinning combined with a fuselage of plastic-bonded laminated birch. Two

Left: **Tsybin Ts-1.**

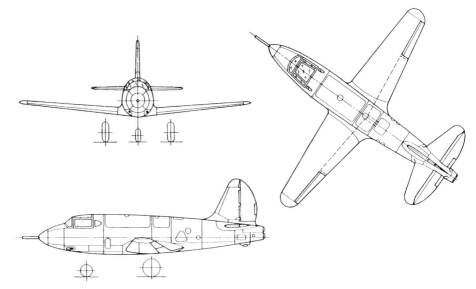

23mm cannon were housed in the lower fuselage and the estimated performance figures indicated that the aircraft would take just one minute to reach 5,000m (16,404ft) and its service ceiling would be 16,000m (52,493ft). Polikarpov died suddenly on 30th July 1944 and, surprisingly, all work on the Malyutka ceased after his loss.

Tsybin Ts-1 and LL-series

Pavel Tsybin will be most remembered for his RS-series of bomber and reconnaissance aircraft described in the companion volume on bombers, but for a short period, under official orders, he also dabbled in rocket aircraft design. In September 1945 MAP's Flight Research Institute (LII-MAP) asked the Tsybin Design Bureau to undertake some research into wing configurations that would be suited for flight at speeds approaching, and possibly reaching, the speed of sound. Following wind tunnel testing held at TsAGI in 1946, work began on a rocket-powered research aircraft called the Ts-1 and this flew in mid-1947. It was towed by a Tupolev Tu-2 to heights between 5,000m and 7,000m (16,404ft and 22,966ft) and, after casting off, the Ts-1 would glide away and enter a dive. On reaching the maximum gliding speed it would then level out and fire its 14.7kN (3,305lb) Kartukov PRD-1500 solid fuel rocket motor. Incidentally, this was the engine's thrust rating at sea level; it gave more at height.

Initially the Ts-1 had a straight wing but, between them, the two airframes to be completed also tested a swept-back wing and a wing swept forward 30° on the leading edge. In this form the aircraft achieved a maximum speed of 1,200km/h (746mph) but both Ts-1s generated a tremendous volume of data and information. They used a take-off trolley and landed on a skid and were also known as the LL-1 and LL-3 (Flying Laboratory Numbers 1 and 3).

'German' OKB Studies

The section above has described the Soviet Union's initial studies into rocket fighter design but a substantial design and research effort was also forthcoming from an OKB that was staffed primarily by Germans. The organ-

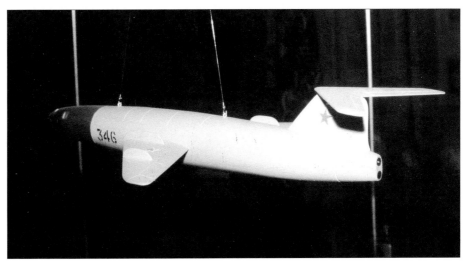

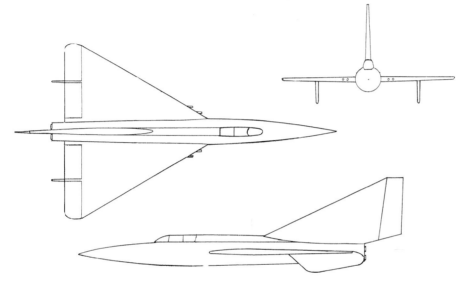

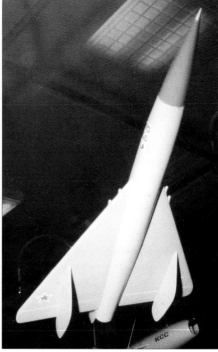

isation of former German designers and engineers into two new Design Bureaux has been described in Chapter Two in the section dealing with the Baade OKB-1. OKB-2, led by Hans Rössing, was formed on 22nd October 1946 and was to be employed on the design of bombers and rocket-powered fighters.

Type 346

OKB-2's first job was to continue working on a proposed supersonic rocket-powered swept-wing research aircraft called the DFS 346 (DFS was actually the former German Institute of Gliding at Griesheim). In 1947 an unpowered 346 airframe (called 346P – P for *planer* or glider) was test dropped from a captured former American B-29 bomber and then in September 1949 another airframe called the 346D, complete with rocket motor, was also test-dropped but again flew as a glider. However, when it landed the second aircraft was badly damaged in an accident and another year was to pass before it became airborne again. That was in October

1950 when it flew for the first time using its own power. On 13th August 1951 the last 346 to be built had its engines fired during its initial test flight, but on 14th September this machine was lost during its third sortie and the programme was closed down.

The Type 346 featured a prone pilot position and had a Walter HWK 109-509C rocket motor (which the Soviets redesignated ZhRD-109-510). This German-designed engine used methanol/hydrazine hydrate (C-Stoff) and concentrated hydrogen peroxide (T-Stoff) to provide instantaneous ignition and was fitted with two thrust chambers. At sea level the combined thrust was 19.6kN (4,410 lb), with 16.7kN (3,750 lb) from the primary chamber and another 2.9kN (660 lb) from the second, which rose to a maximum of about 22.0kN (4,960 lb) at height. The main chamber was fired only for take-off or when maximum thrust was called for but the secondary chamber operated all of the time. Despite having an estimated top speed of Mach 2 (2,127km/h or 1,322mph) at altitude (following a two-minute

rocket burn at full power) and notwithstanding the very large volume of effort that went into this programme, the 346 never achieved even Mach 1. In all two powered airframes were built together with the 346P and also the 346A high-speed glider.

Type 468 and 466

Rocket motors were a relatively simple form of propulsion to operate although their fuels could be quite nasty chemicals and were difficult, and sometimes dangerous, to work with. Nevertheless such engines offered a potentially enormous rate of climb to an interceptor fighter and such a capability could not be ignored. OKB-2's design team therefore began looking at new ideas for an interceptor and the result was the single-seat Type 468 started in 1949. This was a delta-winged aircraft that had 60° of sweep on the leading edge, a huge single swept fin and two more fins on the underside of the wing that might possibly have doubled as skids for landing. It looks as if this aircraft, like many rocket fighters, would have

taken off using a jettisonable dolly. The identity of the rocket motor is unknown but it would have had four thrust chambers. In cross section the fuselage was circular in shape and in the cockpit the pilot was seated in a normal position, the experience of the Type 346's prone pilot arrangement having not been repeated. The result was a very sleek and attractive aeroplane that was to carry two 23mm cannon in each wing root, although an alternative armament of six 132mm (5in) rocket projectiles was also available. Estimated top speed was 2,000km/h (1,243mph) and service ceiling 22,000m (72,178ft).

By now the OKB's team leader Hans Rössing had working alongside him a Russian counterpart called Alexander Beresnjak, together with a well-known engineer from the former Heinkel company called Siegfried Günter. In addition the job of calculating the airframe's structural strength was placed in the hands of Alexej Jelezkij. A full-size wooden mock-up of the 468 was built that would also serve as a model aircraft to test the fighter's flight characteristics and handling. In the same way that the 346P had paved the way for its powered sister aircraft, this full-scale model would be towed to altitude and then released for a series of free-flight trials. The glider, known as the Type 466, was expected to reach a maximum speed of 500km/h (311mph) and in appearance it looked very similar to the 468. It was built in 1950 and tested in the full-size wind tunnel operated by TsAGI, but it was never flown because very suddenly, in June 1951, OKB-2 was closed and disbanded. Work on the 468 was brought to an end and many of the Bureau's engineers were transferred to rocket work.

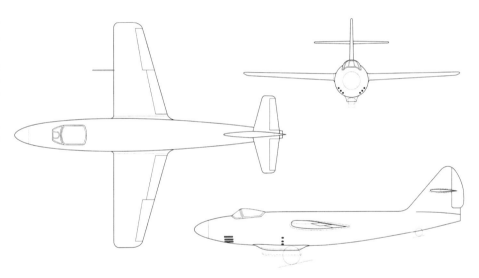

Other Post-War Studies

Such was the interest in rocket power that it is hardly surprising that the major fighter specialist Design Bureaux also had a go. In fact, the Soviet Union probably undertook more research into rocket aircraft design than any other country. What is a little surprising however, is that only a single all-new rocket fighter design was actually built and tested by the 'big four' of Lavochkin, Mikoyan, Sukhoi

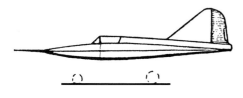

Sketch drawing of the Moskalyov SAM-29, which would have had a wing shaped like a smoothing iron.

and Yakovlev, although in due course several conventional jet-powered types were to have auxiliary rocket motors fitted to them.

Lavochkin 'Aircraft 162'

The is the only known Lavochkin fighter project to have been powered by a rocket and, like many such designs, it was intended to undertake point defence duties close to important military and civil establishments and installations. Work on the design began in 1946 and the aircraft was to be powered by a Dushkin RD-2M-3V twin-chamber liquid-fuelled rocket motor with the exhaust placed at the very end of the fuselage. Cruising at its lowest thrust setting this motor could run for a maximum of twenty minutes when the aircraft was flying at a height of 3,000m (9,843ft). 'Aircraft 162' had a straight wing and tailplane but the fin was swept. Its armament, a battery of six Pobyedonostsev TRS-82 spin-stabilised rocket projectiles, was mounted around the lower nose in two arcs of three. Take-off would be made using a jettisonable two-wheel dolly and during the early stages of the flight the fighter would be guided to its target by radio control. Close in, the enemy would then be attacked using the aircraft's radar and a landing would be completed using the underfuselage skid and fixed tailwheel.

The specified performance figures showed that the '162' was expected to reach 5,000m (16,404ft) altitude in just two and a half minutes and have a service ceiling of

18,000m (59,055ft). Data and material from OKB-2's swept-wing Type 346 project (above) was made available to the Lavochkin Design Bureau during the development programme and a full size mock-up of the aircraft was built. However, the project progressed no further.

Moskalyov SAM-29 (RM-1)

In 1945 both the Mikoyan OKB and the OKB of Aleksandr S Moskalyov received contracts to build a pair of prototype rocket fighters. Moskalyov's studies were collected under the designation SAM-29 (RM-1) and they shared the all-wing Arrow or Strela planform employed on some of his team's earlier piston-powered designs. The Strela was a Gothic delta wing blended into a needle nose and it did not have a horizontal tail. Only sketches of the project have been found but it was to have had a large fin, a Dushkin rocket motor and two nose cannon. The end of the war also brought an end to the project and Moskalyov's OKB was closed in January 1946.

Mikoyan I-270

This was the only all-new rocket-powered design from the principal fighter OKBs to actually get as far as undergoing something of a flight test programme, and that was only brief. Having seen examples of the wartime German Messerschmitt Me 163 and Me 263 rocket-powered interceptors, in 1944 Mikoyan began looking at its own versions of these aircraft; however, it was the 1945 contract for two prototypes that really got things moving. The official designation was I-270, although the aircraft was also called 'Izdeliye Zh', and it featured a straight wing, a T-tail and a tricycle undercarriage. Power came from a Dushkin and Glushko RD-2M-3V rocket that used red fuming nitric acid and kerosene for its fuel with hydrogen peroxide acting as an igniter. It had a main thrust chamber giving 14.2kN (3,200lb) of thrust at sea level and an auxiliary chamber

providing another 3.9kN (880lb), these figures rising by about 15% to 20% at height. Both were needed for take-off and climb but the main chamber would be shut down in flight, and this gave a maximum endurance of nine minutes. I-270 took 2.37 minutes to reach 10,000m (32,808ft) and had a service ceiling of 17,000m (55,774ft). Two 23mm cannon were housed in the lower fuselage.

The first flight specimen was ready before its powerplant and so, in December 1946, several gliding tests were made to check the aircraft's handling. These were undertaken after a Tupolev Tu-2 tug aircraft had pulled the I-270 up to a suitable height and on release the pilot was able to reach a maximum speed of 300km/h (186mph). Both prototypes flew, the second in January 1947 under its own power, but within just a few weeks both airframes had been damaged in landing accidents and they were never repaired. There was really no point because, from a military point of view, the I-270 no longer had any value. This was also the principal reason why Moskalyov's project was terminated.

The knowledge gained from all of this work would never actually lead to an in-service fighter aircraft. However, during the 1950s guided and unguided missiles powered by rocket motors were added to the inventory and these did make use of the experience now available; it was also put to good use in the Soviet space programme. Rocket-powered aeroplanes themselves had some drawbacks, in particular a very short range and endurance that was not always popular with officials. The USSR covered a very large area of land and sea and any defending aircraft that had only a short-range capability would not, in most cases, be very helpful or useful. In due course jet engines were designed which were more powerful and offered the climb and speed performance that the armed forces wanted; they also became more fuel-efficient and so, within a few years, fighter aircraft with a good range were also becoming available.

However during the 1950s, to help make up any shortfall in engine thrust, some conventionally-powered fighters were fitted with auxiliary rocket motors to provide extra performance at high altitude, or to shorten their take-off distance or increase their rate of climb. For example the SM-30 variant of the Mikoyan MiG-19 (Chapter Three) had a very powerful rocket motor fitted beneath the

Mikoyan I-270 prototype.

A Mikoyan SM-30 prototype is seen sitting on its special Zero-Length Launch pad.

The second Mikoyan Ye-50 prototype with its rocket motor in operation.

lower rear fuselage to allow the aeroplane to make a ZELL (zero-length) launch off a special pad. Trial examples of Sukhoi's Su-7 were also fitted with RATO units to reduce their take-off distance.

Mikoyan Ye-50 and Ye-50A

This was a version of the swept wing Ye-1 prototype (Chapter Three) that was intended to operate as a mixed-power interceptor (in fact it was known initially as the Ye-1A). It had an RD-9Ye jet engine in the usual position in the rear fuselage, which gave 37.2kN (8,375 lb) of thrust with afterburning (the planned RD-11 unit was not yet ready), and a Dushkin S-155 rocket motor fitted above the main engine afterburner. This unit supplied another 15.7kN (3,525 lb) of thrust at altitude (rocket thrust increases with height) and used kerosene as its main fuel with nitric acid as an oxidant. It was only to be fired during an interception, the jet operating alone during the rest of the flight including the take-off and landing. The first Ye-50 prototype made its maiden flight on 9th January 1956 but the first in-flight firing of the rocket did not take place until 8th June. However, the presence of this second engine also took up much of the space previously available for jet fuel, so its endurance again proved to be insufficient; the rocket also increased the empty weight by 714kg (1,574 lb). The Ye-50 recorded a maximum speed of 2,460km/h (1,529mph) or Mach 2.32 at 18,000m (59,055ft) and a service ceiling of 23,000m (75,459ft).

Two more prototypes were constructed and in 1956 the Ye-50A was proposed which was to receive the AM-11 jet and have all of its rocket fuel stored in a large ventral tank; as a consequence more room would also be available for conventional fuel. A mock-up was built, one prototype was planned and two more series aeroplanes were to follow, but then in 1957 the Ministry decided that even more powerful jet engines would be the best way forward and so it stopped all work on mixed-power fighters. On Ministry orders the Dushkin rocket Bureau was closed and further work on the Ye-50/50A programme was cancelled.

FLYING BOAT FIGHTERS

In the companion volume on bombers a number of pages were devoted to the Soviet Union's efforts to produce a jet-powered flying boat bomber – some of the results were truly remarkable designs. However, like most of the rest of the world, the USSR appears to have put relatively little effort into producing a flying boat fighter. The greatest exponent of advanced flying boat fighter design was probably Convair in America, who designed a series of such aircraft and finally flew a proto-

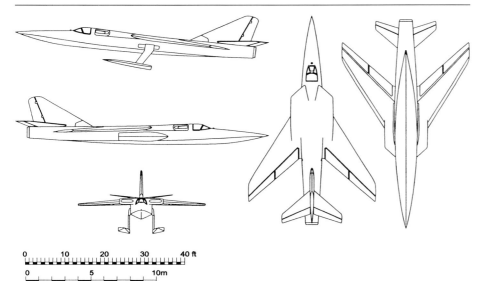

0   10   20   30   40 ft

0   5   10m

TsAGI Design 4221 with twin hydroskis (1955).
Jens Baganz

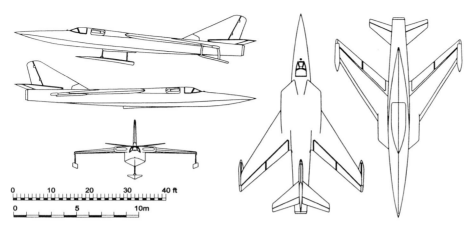

clear of the water so allowing it to skim the surface until take-off. The bonus was that once retracted into the body, these skis offered much less drag than either buoyant floats or a flying-boat hull. TsAGI's experiments confirmed that retractable skis could be used for operations off water.

The two projects studied, Designs 4221 and 4222, were similar. Their layouts had a span of 10.0m (32ft 9½in), length 19.2m (63ft 0in), wing area 40.0m² (430.1ft²) and an estimated take-off weight of 10,000kg (22,046 lb); the wings were swept 60°. Results showed that the ideal solution was to use two skis along the sides of the fuselage, although Design 4221A had a single main hydroski under the body and two supporting skis on cantilever legs fitted to the wingtips, which made the problem of retracting the skis after leaving the water rather simpler. Stability on the water was good and the take-off speed was 250km/h (155mph). Despite this promising research, as already noted none of the main fighter OKBs took up the challenge of designing a boat fighter. Perhaps the fact that Mikoyan, Sukhoi and Yakovlev were all based in Moscow, some distance from the sea, could have been a factor.

TsAGI Design 4221A with a single main hydroski and two smaller supporting outrigger skis (1955).
Jens Baganz

type in 1953 that would show supersonic performance, the XF2Y-1 Sea Dart. It appears that the main Soviet OKBs did little or no research into this type of aircraft (at least no documents have been found to date) but the TsAGI Research Institute certainly did.

TsAGI Projects 4221 and 4222

During 1955 the Central State Aerodynamic and Hydrodynamic Institute at Zhukovskiy (TsAGI) studied two designs for a hypothetical jet fighter that used retractable hydroskis to get off the water. Taking-off from water in a high-speed aeroplane was possible using a ski or skis which extended under the surface like seaplane floats and then, as the aircraft's speed increased, pushed the fuselage upwards

Rocket-Powered Fighters and Research Aircraft – Data / Estimated Data

Project	Span m (ft in)	Length m (ft in)	Gross Wing Area m² (ft²)	All-Up-Weight kg (lb)	Engine kN (lb)	Max Speed / Height km/h (mph) / m (ft)	Armament
Bereznyak-Isayev BI 3rd Prototype (flown)	6.6 (21 8)	6.935 (22 9)	7.2 (77.4)	1,650 (3,638) (Higher on later a/c)	1 x D-1A rocket 12.7 (2,865) at height	900 (559)	2 x 20mm cannon (Not fitted)
Florov 4302 (flown)	6.932 (22 9)	7.124 (23 4.5) 2nd a/c, 7.152 (23 5.5) 3rd a/c	8.85 (95.2)	1,750 (3,858) 3rd a/c	1 x rocket 10.7 (2,425)	826 (513)	None
Kostikov 302	11.4 (37 5)	8.708 (28 7) without guns	17.8 (191.4)	c3,800 (8,377)	1 x D-1A rocket 12.7 (2,865) at height + 2 ramjets	Estimated 800 (497) at S/L, 900 (559) at height	2 x 20mm cannon
Kostikov 302P (flown)	9.55 (31 4)	8.708 (28 7) without guns	14.8 (159.1)	3,358 (7,403)	1 x D-1A rocket 12.7 (2,865) at height	?	2 x 20mm cannon
Polikarpov Malyutka	7.5 (24 7)	7.3 (23 11.5)	8.0 (86.0)	2,795 (6,162)	1 x NII-1 9.8 (2,205)	890 (553) at S/L	2 x 23mm cannon
Tsybin Ts-1 (LL-1) (flown)	7.1 (23 3.5)	8.98 (29 5.5)	10.0 (107.5)	2,039 (4,495)	1 x PRD-1500 14.7 (3,305)	1,050 (653) at height	None
Tsybin Ts-1 (LL-3) (flown)	7.22 (23 8) forward sweep	8.98 (29 5.5)	10.0 (107.5)	2,039 (4,495)	1 x PRD-1500 14.7 (3,305)	1,200 (746) at height	None
OKB-2 Type 346 (flown)	9.0 (29 6.5)	15.987 (52 5.5) with pitot	14.87 (159.9)	5,230 (11,530)	1 x ZhRD-109-510 22.0 (4,960) at height	Estimated 2,127 (1,322) at height	None
OKB-2 Type 468	c8.0 (26 3)	c15.0 (49 2.5)	?	?	1 rocket motor	2,000 (1,243) at height	4 x 23mm cannon
Lavochkin '162'	8.96 (29 5)	11.175 (36 8)	?	5,000 (11,023)	1 x RD-2M-3V c21.75 (4,895) at height	1,050 (653) at S/L, 1,100 (684) Mach 0.962 at 5,000 (16,404)	6 x TRS-82 rocket projectiles
Mikoyan I-270 (flown)	7.75 (25 5)	8.915 (29 3)	12.0 (129.0)	4,120 (9,083)	1 x RD-2M-3V c21.75 (4,895) at height	1,000 (622) at S/L, 936 (582) at 15,000 (49,213)	2 x 23mm cannon

Rocket Power and Flying Boats

Chapter Eight

Variable Geometry and Fast Interceptors

SWING WING PROJECTS

During the 1960s the Soviet fighter OKBs managed to shake off the effects of Khrushchev's missile theories and move forward in two principal directions. One of these was the development of a very fast heavy interceptor that led to the Mikoyan MiG-25 described in the second half of this chapter. The second matched developments in other countries with the adoption of variable-geometry (VG) swing wings. Variable-sweep wings offered, within limits, an optimum sweep angle for a single aeroplane wishing to fly at different Mach numbers. Low sweep angles are needed to give a good subsonic performance and to reduce take-off and landing distances to a minimum while high speeds and supersonic performance will benefit from having more sweep. The ability to combine these

qualities within a single airframe was an exciting development and swing-wing designs were produced in several countries including America, France and the UK. However, besides its swing-wing proposals the line of research undertaken at the Mikoyan Design Bureau also embraced some more vertical take-off designs.

Mikoyan

Mikoyan MiG-23M

The need for a new fighter-interceptor to replace the MiG-21 in the Soviet Union's Frontal Aviation (VVS) forces was acknowledged by an official resolution issued on 3rd December 1963 and a prototype was to be ready for state testing in late 1965. It is currently unknown if Yakovlev or Sukhoi

One of the two all-new Tupolev 'Aircraft 138' configurations had a canard and cranked delta wing. This model also shows the stowage position for the four K-80 missiles.

responded to this requirement with designs of their own (although Sukhoi was by now looking at developments of the Su-7 with VG wings which became the Su-17 below) but Mikoyan put two separate design teams onto the job. One examined variable-geometry wings in a project that led to the MiG-23 described shortly, while the second looked at fixed delta wings but combined with a battery of lift engines in the centre fuselage to give STOL performance. The idea was to determine which configuration best suited the needs of the Air Force but the new aircraft also needed a good radar and thus a large radome. Consequently, for the first time at the

were two 23.0kN (5,180lb) Kolyesov single-shaft RD-36-35 lift jets located in the mid-fuselage bay behind the pilot. These had already been used on several V/STOL aircraft described in Chapter Six and here they were set forward at an angle of 85°. Just like the 23-31 prototype the 23-01 had rectangular louvred 'doors' on top of the fuselage to allow air to get to the lift engines for take-off and landing. Underneath the fuselage both thrust nozzles had a rotating grid of seven deflector vanes to enable the pilot to vector the thrust through a limited angle, which could augment the main engine for take-off and also act as a thrust reverser for landing.

The prototype 23-01 did not actually receive any radar and so some ballast had to be loaded in its place to compensate for the reduction in weight. The air intakes were set a little away from the fuselage to separate the boundary layer and included a movable half cone. A non-operational GSh-23L twin barrel 23mm cannon was fitted beneath the fuselage and underwing pylons held one Vympel K-23R and one K-23T air-to-air missile, although only dummy weapons could be carried because there was no fire-control equipment in place. Vano A Mikoyan (the brother of the Bureau's general designer) was given overall charge of both the VG and STOL programmes but A Andreyev led the 23-01's design team.

The Mikoyan 23-01 made its first flight on 3rd April 1967 but by then it was clear that the swing-wing variant, the prototype of which was now called the 23-11, would be the better choice. In fact chief designer Artyom Mikoyan had already decided in favour of the variable-geometry wing and, as a result, the flight-testing undertaken on the 23-01 was minimal with no attempt being made to determine a full flight envelope; in fact the test programme only got as far as establishing the parameters for take-off and landing. One of the reasons for this was that there were immediate problems with the STOL mixed cruise and lift engine arrangement as a whole. What the flight tests revealed was some instability and loss of control during take-off and landing and the weakness was particularly apparent during the latter because of an unpredicted suction effect caused by gases from the lift engines. Thanks to this the 23-01 was never flown at speeds of less than 150km/h (93mph) and, in all, it is

Mikoyan OKB, side intakes would also have to be employed rather than the previously favoured nose inlet.

In fact Mikoyan had no experience with lateral inlets and so it was necessary to determine the design parameters required for this type of intake. The first model shows a design proposal called the MiG-23M which had flat side intakes designed along the lines of the VG MiG-23 as built, plus underwing air-to-air missiles on pylons and some underfuselage weapon stations. This project also used the tail assembly from the Ye-8 prototype (Chapter Four) but the nose was too small to hold a radar. The use of the MiG-23M designation here was a consequence of the Ye-8 having previously received the provisional title of MiG-23 – the designation MiG-23M was later used for a major production variant of the VG MiG-23 itself. An alternative model fitted with lift engines, and with MiG-23 written on the

side, was similar to the MiG-23M model but had semi-circular intakes and a larger nose to hold a radar. After refinement this design was turned into a prototype called the 23-01. No data is available for any of the designs produced in model form, but one assumes that dimensionally they would have been pretty close to the 23-01.

Mikoyan 23-01

Both STOL and VG designs were to receive the new Khachaturov (Tumanskiy) R-27F-300 engine as their main power unit and, in due course, a Sapfir-23 (Sapphire) radar; the main armament was to be four medium-range air-to-air missiles. The R-27F-300 had a maximum dry thrust of 51.0kN (11,465lb), increasing to 83.3kN (18,745lb) in reheat, and Mikoyan studies to find a home for this excellent engine had actually started way back in 1960. In addition for the STOL project there

Above and right: **'MiG-23' STOL fighter model with semi-circular side intakes. Again the dorsal engine doors are visible.**

Bottom: **This model is, in appearance, close to the Mikoyan 23-01 as built, but it does have a smaller nose radome.**

believed that only fourteen flights were actually completed.

Nevertheless, sufficient test flying was recorded to allow the 23-01 to take part in the Domodedovo air show held on 9th July 1967, which resulted in the fighter being given the ASCC codename *Faithless*. Its display was brief so it seems likely that participation in the event was undertaken simply to try and give the West an impression that this was a new service type. The 23-01 was also designated MiG-23PD (PD for *podyomnye dvigatyeli* or lift engines) to cover any likelihood of service with the VVS. Following the show 23-01 joined the prototype 23-31 at the Moscow Aviation Institute where it served as an instruction aid for aeronautical students. Today the 23-01 prototype no longer exists and, to date, no accurate estimated performance figures have ever been released.

Variable Geometry and Fast Interceptors

Mikoyan 23-11 and MiG-23

As noted, by the time the 23-01 flew, Mikoyan's designers were confident that an aircraft with a variable-geometry wing was the best way to fulfil the new requirement. Nevertheless, it was appreciated that this too was a new and relatively unknown aerodynamic feature and some big problems would have to be dealt with before a prototype aeroplane could be put together. To help the Design Bureau the Ministry of the Aircraft Industry ordered other establishments and companies to give Mikoyan all possible assistance in refining the new technology and the

techniques that would be required to use it.

However, some good fortune was forthcoming in the form of the American General Dynamics F-111A strike aircraft that made its first flight on 21st December 1964. Since design work on the VG-wing 23-11 prototype did not start in earnest until 1965, the Mikoyan team was able to use some of the American solutions to the technical problems it was experiencing. The most obvious was the wing pivot position and, after this had been observed on the F-111, it was possible to design suitable inner wing panels (or gloves) to go with it (a smooth airflow over the com-

plete installation was essential for the stability of the aircraft at high angles of attack and low speeds). Also, like the F-111, the designers chose a sweep angle of 16° for take-off and landing, 45° for flight at cruise speeds and 72° for supersonic speeds. Former test pilot Gregoriy Sedov headed the design team working on this new fighter.

There were delays with the planned R-27F2-300 engine, a development of the R-27F-300 used in the 23-01, and in due course one of the older units had to be installed in the prototype to keep the programme up to speed. It is worth noting that the engine's thrust rating figures were deliberately pitched lower than their possible maximum in an attempt to avoid the criticisms of unreliability that had been made against all of Mikoyan's previous supersonic aeroplanes (this step proved fairly successful). Other differences from the 23-01 included rectangular intakes with a large splitter plate (instead of semi-circular intakes) while the planned armament was four K-13 or K-23 air-to-air missiles, one under each wing and two under the fuselage. Again no radar was installed, only ballast, and

The Mikoyan 23-01 prototype with dummy missiles aboard.

Mikoyan 23-11 first prototype with wings in the maximum sweep position.

Soviet Secret Projects: Fighters Since 1945

Mikoyan had to get official permission to build a series of 23-11 development aircraft to test the swing-wing mechanism thoroughly.

The first 23-11 prototype made its first flight on 10th June 1967. On the second flight the variable-geometry wing was tested in all positions, on the third trip the aircraft went supersonic and on 9th July 1967, the pilot gave a brilliant performance at Domodedovo; in due course the customary Western codename, *Flogger,* was given to the new type. The 23-11 had a service ceiling of 17,200m (56,430ft) and, later on, this airframe was used to test two new engines – the 88.2kN (19,850lb) R-44 and the 98.0kN (22,050lb) R-47 (thrust ratings which equate to operation in reheat). Retirement to the Air Force Museum at Monino followed.

The 23-11 did indeed form the basis for the new production fighter and at last, after several false starts, the Soviet Air Force finally had a MiG-23. Initially the objective was to produce an aircraft that was purely a high-speed interceptor, with little thought being given to turning air combat. The first production aeroplane was flown on 21st May 1969 and, as one might expect with such an advance in aerodynamics, there were development problems. After several 'interim' versions had been produced, the definitive aircraft (designated MiG-23M) entered production in 1972, by which time the type had matured into an altogether more capable air defence fighter with increased manoeuvrability and a more powerful engine. It proved to be a great success and was built in several versions, some of which are still in service, and with much more in the way of avionics on board it was a far more sophisticated air-

A Sukhoi Su-17 loaded up with four PTB-600 drop tanks.

craft than the earlier MiG-21. Many examples were built for overseas services. On 9th December 1970, while the MiG-23 was being prepared for service, Artyom Mikoyan died. He was replaced as the head of the Design Bureau by Rostislav A Belyakov.

Mikoyan MiG-27

American experience in the Vietnam War showed that another important type for the inventory would be a tactical strike aircraft. Eventually a version of the MiG-23 was developed as such, which is described in the companion volume on bombers. Suffice to say that it began life as the MiG-23B but later became the MiG-27. The main changes included new ground-attack equipment and a different nose plus a new engine, but the *Flogger* codename was retained by the ASCC. The first prototype flew on 20th August 1970.

Mikoyan MiG-23A/MiG-23K

A naval version of the MiG-23 was also proposed which was based on the lighter and improved MiG-23ML variant produced in the mid-1970s. The naval aircraft was initially called MiG-23A and was intended to be a multi-role aircraft produced in fighter, bomber, attack and reconnaissance forms and powered by an R-29-300 engine. The main differences from the Air Force aircraft included an improved view from the cockpit for carrier operations, a stronger undercarriage, a landing hook and the replacement of the single folding ventral fin by twin strakes. The type was to be operated by the planned Project 1160 aircraft carriers, which would have been fitted with catapult take-off and arrester equipment. The MiG-23A was proposed in 1972 but the large Project 1160 was abandoned and the aircraft died with it.

Side view of the Mikoyan MiG-23K naval fighter proposal (1977). Russian Aviation Research Trust

Five years later the MiG-23K (32-31) carrier-based fighter was offered, again based on the MiG-23ML but also taking in knowledge gained from the MiG-23A. It was to be powered by an R-100 bypass engine and would go aboard the proposed Project 1153 conventionally-powered aircraft carriers. The MiG-23K differed from the MiG-23A in having a wing of increased area and fitted with double-slotted flaps, an even larger cockpit canopy and the capability to perform in-flight refuelling. Again this version was not built and the Soviet Navy's aviation aspirations moved on to the Sukhoi Su-27K, Mikoyan MiG-29K and Yakovlev Yak-41. The weaponry intended to be carried by the MiG-23K included air-to-air and air-to-surface missiles.

Sukhoi

Sukhoi Su-17 and Su-22

Sukhoi's Su-7 tactical fighter-bomber (Chapter Three) proved to be a major success but the appearance of the American F-111 programme inspired some other aircraft designers to take a look at VG wings. P P Krasil'shchikov of TsAGI approached Sukhoi with the idea of fitting new variable-sweep outer wing sections to the Su-7B, with the

inner wing still fixed. As a result, during 1963 extensive wind tunnel testing was carried out by TsAGI on this type of flying surface and in 1965 N G Zyrin was made Chief Designer for the overall project, which was to embrace fitting VG wings to an Su-7BM airframe. The resulting aircraft was designated S-221 and its old 63° fixed-sweep wing was replaced by one with a pivot point well away from the fuselage; the minimum sweep angle for the new outer wing was 30° and the maximum possible sweep was 63°.

The Su-7BM was rebuilt with its new wings during the first half of 1966 and made its first flight on 2nd August; flight-testing gave good results. The new surfaces, with their extra mechanism and complexity, increased the structure weight while also taking away some of the space available for fuel, but cruise flight with the wings in the forward position actually revealed an increase in flight endurance. As a result, although essentially a research aircraft, orders were given to put the type into production and this was given the designation Su-17. It really was a second generation Su-7 and extended the production and service life of Sukhoi's tactical fighter by a considerable period. The production prototype was designated S-32 and the Su-17 replaced the older fixed-wing versions on the line. Series aircraft introduced a large dorsal spine stretching from the canopy to the fin but, apart from this and the new wing, all other changes were of a detail nature. In the West the Su-17 became the *Fitter-B*, the Su-7B being retitled *Fitter-A*. Further versions designated Su-20 and Su-22 followed and these VG types served primarily as ground-attack aeroplanes.

FAST INTERCEPTORS

The second route taken to find a new fighter was the creation of a large aircraft that was intended to be capable of very high speeds. Besides Mikoyan's winning MiG-25, it is known that Tupolev studied some competitive projects under its 'Aircraft 138' designation. One assumes that Sukhoi would have undertaken research into this type of aircraft as well but, to date, no information has come to light to support this. The arrival of the American Convair B-58 Hustler supersonic bomber into operational service in 1960, and the start of the Lockheed SR-71 Blackbird Mach 3 reconnaissance aircraft programme at around the same time, forced the Soviet Union's leadership to take action to create an antidote to these threats. Part of this response was to begin work on new supersonic bomber and cruise missile carrier types designed by the Tupolev and Sukhoi Bureaux, the 'Aircraft 135' programme and the T-4 prototype respectively, which are detailed in the companion volume on bombers. In addition a new interceptor that could deal with the American aircraft was vital and the result was the MiG-25.

In fact the Mikoyan Bureau was eventually to produce a combined multi-role airframe that could undertake interception duties for the PVO and also reconnaissance for the VVS. The first preliminary sketches for designs leading to the MiG-25 were actually drawn in 1958. Very high speed and a comprehensive suite of avionics were the key elements but full design work did not get going until mid-1959. Very quickly it became apparent that the project had great potential, but the very severe flight performance parameters would necessitate a new approach to the design of the avionics and weaponry and also the introduction of new production techniques.

Mikoyan Ye-155

Since the requirements stated by the PVO and the VVS for the interceptor and reconnaissance aircraft were broadly similar, with a top speed of around Mach 3 and a service ceiling in excess of 20,000m (65,617ft), it was

Left and top: **Model of the unbuilt Mikoyan Ye-155R project fitted with variable sweep wings (c1961). These pictures show the wing in the minimum sweep position.**

Above: **The STOL Mikoyan Ye-155R project model showing the lift engine doors in the open position (c1961).**

Soviet Secret Projects: Fighters Since 1945

officially confirmed in 1960 that a single aircraft design could indeed fulfil both roles. In February 1961 the Central Committee of the Communist Party together with the Council of Ministers issued a joint directive ordering the Mikoyan Bureau to develop the aircraft. The interceptor was designated Ye-155P and the reconnaissance version Ye-155R and both the VVS and PVO wrote general operational requirements to cover their versions. The most suitable engine was the R15B-300 designed by A A Mikulin and his closest aide S K Tumanskiy, which was a spin-off of an earlier axial turbojet called the Model 15K that had been proposed for a remotely-piloted aircraft. Very quickly the engine designers altered the compressor, combustion chamber and afterburner, increased the gas temperature throughout the whole engine and incorporated a new variable three-position ejector nozzle. Engines designed by Tumanskiy embraced his principles of robustness and simplicity and the resulting R15B-300 was a large single-spool axial turbojet that had just five stages of compressor blading and a single-stage turbine.

As back-up for the programme, the Ye-150 and Ye-152 family of heavy interceptors described in Chapter Five was on hand to help refine various other items of equipment, including analog computers, communications systems, command link equipment and ejection seats. In addition the high speeds achieved by these aircraft contributed valuable data on aerodynamics, gas dynamics, flight controls and aircraft stability and controllability at high Mach numbers. M I Guryevich was made responsible for the airframe and N Z Matyuk for the integration of equipment and weapons, but Guryevich was suffering from failing health and eventually had to retire, leaving Matyuk with full responsibility for the project.

Early on there were three possible arrangements for the Ye-155, all of them twin-engined due to the aircraft's substantial weight. One had its engines located side-by-side like the Ye-152A and another had a stepped arrangement similar to the Mikoyan I-320 experimental fighter (Chapter Four) with one engine amidships exhausting under the fuselage and another in the rear fuselage. The third project had its engines mounted one above the other in the rear fuselage, in a form that was identical to the British English Electric Lightning interceptor. The second and third options were rejected because the powerful engines had to have a large diameter, which increased the aircraft's height by an unacceptable margin; this aspect also turned the removal of the engines for servicing or replacement into a complicated operation. The idea of placing the engines in underwing nacelles was also turned down because if an engine failed during take-off or landing there would automatically be a high level of thrust asymmetry which would be very dangerous. The designers also did away with the old single nose air intake and circular fuselage, so characteristic of earlier Mikoyan jet fighters, and replaced them with rectangular lateral scoop intakes (with movable ramps for adjusting the airflow); these proved to be very successful.

Tunnel testing at TsAGI found that a trapezoidal-shaped wing with an unswept trailing edge would give an adequate lift/drag ratio at speeds between Mach 2 and 3 and a wing structure that was relatively light and that provided adequate space for fuel. Early projects envisaged a tail unit with a single fin and rudder and canard foreplanes to augment the all-moving stabilisers for pitch control, but these were rejected for twin fins and a low-set tailplane. A development of the Tupolev Tu-128's Smerch radar was selected for the Ye-155P interceptor that offered a detection range of up to 100km (62 miles). However, during the early design stage several more alternative layouts were assessed before the Ye-155/MiG-25's final general arrangement was confirmed.

Mikoyan Ye-155R designs

For one of the preliminary studies for the Ye-155R reconnaissance variant the designers attempted to marry the basic airframe to a variable-geometry swing-wing. At maximum sweepback the wing panels and stabilators came together to form what in effect was a delta wing, which improved the aircraft's speed capabilities, while at minimum sweep the aircraft's manoeuvrability, endurance and particularly its short-field performance were also much improved. This project had two crew with a navigator's station (complete with small rectangular windows) located in the nose ahead of the pilot's cockpit, but this feature, together with the VG wing, increased the aircraft's maximum take-off weight. In truth the Ye-155R was intended solely to undertake photographic reconnaissance at high speeds and high altitudes and so good manoeuvrability would be of little benefit. Eventually it was decided that having a navigator was also unnecessary and so a single-seat type was adopted for the reconnaissance version and this two-seat VG wing design was abandoned. In appearance the VG Ye-155R shared the same large side intakes and twin fins of the eventual MiG-25, and it seems likely that the engines were also the same.

Although this book is dedicated to fighter and interceptor types, it seems appropriate to examine some of these reconnaissance studies that formed part of the overall Ye-155 programme. These projects were very impressive and it would have been entirely logical to have adapted them for the interceptor role, had they been built. A second alternative Ye-155R design had small auxiliary lift engines, examples of the RD-36-35 turbojet developed by the Rybinsk Design Bureau under the control of P A Kolyesov. The RD-36-35s were intended to improve the Ye-155R's field performance and were located on both sides of the mid-fuselage spine in a near vertical attitude. They breathed through upper fuselage intakes. These had their own aft-hinged covers that in cruise flight were closed flush with the fuselage topside. Like the swing-wing Ye-155R,

This is the first Mikoyan Ye-155R reconnaissance prototype which became the first 'MiG-25' to fly.

this STOL project had a crew of two with the navigator again seated in a nose compartment ahead of the pilot.

Apart from the second crew compartment with its small windows, and the lift jets whose introduction may have required the airframe to be lengthened, the STOL Ye-155R differed little in general arrangement and shape to the Ye-155 as built. Once again, however, the lift jets were of little use to the aircraft's reconnaissance mission and, in fact, they reduced its range quite appreciably since they absorbed some of the space formerly available for internal fuel. Therefore, the STOL Ye-155R was also abandoned as being impractical. Another variant, an attack/reconnaissance design called the Ye-155ShR, was a proposed dual-role aircraft optimised for ground attack and low-altitude reconnaissance, but it also never left the drawing board.

Mikoyan Ye-155P and MiG-25

Once the above alternatives had been discarded work progressed on the layout selected for construction. Originally the Ye-155P's armament was to have consisted of two (and later four) of Mikoyan's own K-9M missiles, a version of the K-9 used by the Ye-152 but adapted for operation with the Smerch radar. However, as work on the Ye-155P proceeded, the Design Bureau led by M R Bisnovat proposed giving the new fighter a new missile called the K-40 (which in service became the R-40). This had a titanium body that offered a better level of heat resis-

tance when flying at high Mach numbers and also saved weight. There were to be two variants – one with semi-active radar homing (the R-40R) and a second with infra-red guidance (R-40T). Later versions of the MiG-25 interceptor could also carry the R-60 short-range AAM while the MiG-25s that entered service as reconnaissance/strike aircraft had a variety of bomb loads at their disposal. The interceptor was to operate within the Vozdookh-1 (Air-1) ground control system.

Calculations had shown that a single fin and rudder on the Ye-155 would not provide sufficient directional stability, except if the fin were overly large. Thus twin tails slightly canted outwards were fitted, augmented by ventral strakes on the aft fuselage and small fins at the wingtips. The selection of a suitable structural material that could cope with the kinetic heat created by flight at such high speeds was a singularly important issue. Prolonged operation at such speeds would quickly soften aluminium alloys and the airframe's inherent strength would be lost. One alternative was titanium but this material was difficult to fabricate and current techniques for welding often left the joints susceptible to cracking. Steel was another and this could be welded without difficulty, which also removed the need for special sealants and labour-intensive riveting. The decision was made to use steel but several members of the design team had doubts about employing such a heavy material. For example it was predicted that integrally welded steel fuel

tanks would be unable to absorb the heavy loads that would be experienced during flight and they would eventually crack open. Fortunately, extensive static testing showed that this would not be the case.

In early 1962 a second official directive was issued ordering the construction of several prototypes. It was planned that the reconnaissance Ye-155R should fly first and during 1962 a mock-up of this version passed through its official inspection. By late 1963 the first prototype was largely complete and the Ye-155R made its first flight on 6th March 1964. The first interceptor prototype was completed in the summer of 1964 and made its first flight on 9th September. Ye-155P was basically similar to the Ye-155R but the latter's 'camera' nose gave way to an ogival radome to house the Smerch-A scanner, although on the first Ye-155P the radar set and dish were substituted by test equipment. The first Ye-155P went on to complete most of the manufacturer's trials and also found time to break several world records. On 9th July 1967 three of the development Ye-155Ps and one Ye-155R took part in the Domodedovo fly past, which resulted in the ASCC codename *Foxbat* being given to the aircraft. The interceptor became *Foxbat-A* and what eventually

One of the Ye-155P interceptor prototypes displays a full weapon load of four missiles. It also has the type's distinctive 'webbed feet' vertical wingtip endplates which contained anti-flutter weights and were designed to improve directional stability.

Soviet Secret Projects: Fighters Since 1945

Early studies for Tupolev's 'Aircraft 138' were based on relatively minor improvements to the Tu-128. This model retains the wing nacelles to house the main undercarriage legs, a common feature on Tupolev aircraft from this period.

became a combined reconnaissance/strike version was called *Foxbat-B*.

On 28th April 1970 the interceptor passed its official state trials and was cleared for production as the MiG-25P. Series manufacture began in 1971, by which time several changes had been introduced to improve the aerodynamics and stability. These included enlarged and recontoured fins and ventral strakes, increased wing incidence, the triangular 'webbed feet' endplate fins were deleted and differentially movable tailplanes had been introduced. As one might expect with such an advanced aeroplane there were development problems and some fatal crashes. For example wingtip deflection during 5G manoeuvres at very high speeds resulted in aileron reversal and a subsequent loss of control, so the type's maximum speed was eventually limited to Mach 2.83 (the alternative would have been to redesign the wing). In the end almost 1,200 MiG-25s of all versions were built. The MiG-25P had a service ceiling of 20,700m (67,913ft) and from rest could climb to 20,000m (65,617ft) and reach a speed of Mach 2.35 in 8.9 minutes.

The MiG-25 was a landmark design and a phenomenal achievement. Capable of approaching three times the speed of sound, it was built of welded steel in a way that was never adopted in the West. Designed just prior to the MiG-25 and flown in 1962, a British Mach 2 research aircraft called the Bristol 188 was built in welded stainless steel but it proved a nightmare to construct and remained a prototype. In 1963 the American Lockheed company flew the prototype of its YF-12A interceptor, a variant of the SR-71 Blackbird

reconnaissance aircraft. This aircraft was designed to cruise at Mach 3 and over 90% of its structure was made of titanium alloy, another difficult material to fabricate. The number of very high-speed aeroplanes built of metals other than aluminium light alloy are few in number, which makes the manufacture of the MiG-25 in such quantity all the more remarkable. However, the Soviet type also represents something of a paradox because inside that very advanced airframe were electronics based on vacuum tube technology rather than microchips, equipment that was becoming very out of date. Nevertheless the

appearance of the MiG-25 caused a big shock wave of worry throughout the Western nations because it presented a difficult and complex aircraft to counter.

Tupolev 'Aircraft 138'

The known competition to the MiG-25 was this series of projects also aimed at producing a long-range supersonic interceptor fighter. During the 1960s the Tupolev Design Bureau continued to work on a new missile-armed aircraft for the PVO as a follow-on to the Tu-128 (Chapter Five). This effort started as the 'Aircraft 138' project launched in 1962 and

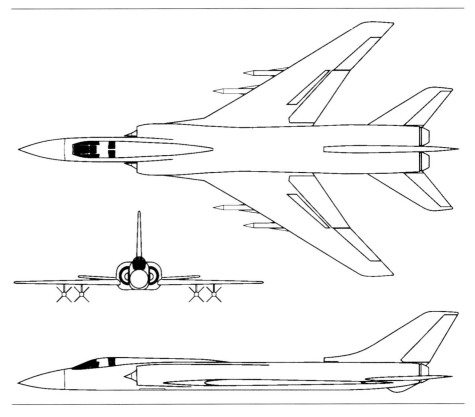

Tupolev 'Aircraft 138' based on the Tu-128 (1963/64).

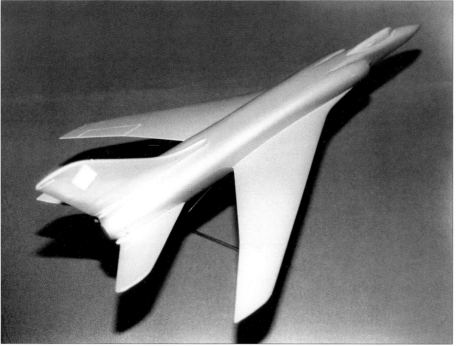

This model represents the Tupolev 'Aircraft 138' based on the Tu-128 and would have been the common airframe for the Tu-138-60 and Tu-138-100.

was initially classed as a continuation of the Tu-28A, but eventually it embraced three different lines of research. One was a modest upgrade of the Tu-128, the second embraced the Tu-138-60 and Tu-138-100 interceptors with a new wing (this received the most attention) and finally there was a pair of all-new designs.

In 1963 and 1964 two proposals were prepared with the designations Tu-138-60 and Tu-138-100, both of which were based on the same 'Aircraft 138' interceptor airframe but they had different radars and missiles. As such they were expected to offer a much improved intercept capability against high-speed enemy aircraft. In this form 'Aircraft

138' retained the basic layout of the Tu-128 but had more powerful Dobrynin VD-19 engines and improved aerodynamics. The VD-19 was produced by the Rybinsk OKB under the control of Vladimir Dobrynin and offered the capability for the '138' to exceed 2,000km/h (1,243mph). Work actually began on this power unit in the autumn of 1958 and the first example was bench-tested towards the end of 1960.

The aerodynamic improvement was to be achieved by the introduction of a new clean wing of reduced thickness. In addition, unlike the Tu-128, this design had its main undercarriage legs and wheels housed in the wing roots and also, in part, in the fuselage, which allowed the designers to do away with the wing nacelles. Tu-138-60 was to have a Smerch-A radar and K-60 air-to-air missiles while the Tu-138-100 would receive a Groza-100 radar and longer range K-100 missiles. These improvements were intended to increase the potential of the fighter to intercept high-speed targets and its maximum speed with the missiles aboard was expected to be much higher than the Tu-128. In addition the patrol endurance would be extended, take-off and landing characteris-

Above left and right: **One of the two all-new Tupolev 'Aircraft 138' configurations had a canard and cranked delta wing. This model also shows the stowage position for the four K-80 missiles.**

Right: **All-new version of the Tupolev 'Aircraft 138' (1963).**

Bottom: **The second new Tupolev '138' layout had a large delta wing only, but could still carry four AAMs.**

tics enhanced and the radar detection and acquisition distances improved; the attack range of the AAMs was also increased.

Research into 'Aircraft 138' reached an advanced stage and included a considerable amount of wind tunnel testing at TsAGI, but these tests revealed some shortcomings with the wing and the overall design. It was found that the required aerodynamic efficiency (lift to drag ratio) at subsonic cruise speeds could not be achieved because of the significant increase in fuselage cross-section brought about by the wider VD-19 engines and the larger diameter of the new radars. Consequently, the specified range and endurance were unlikely to be achieved. In addition the

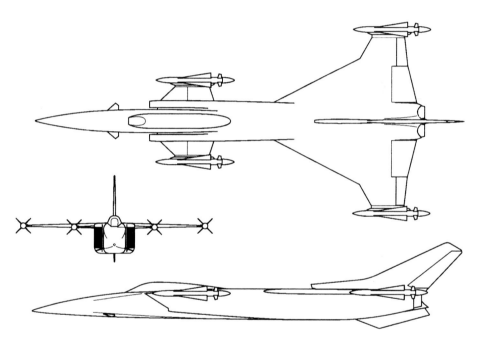

take-off and landing characteristics of the new wing were found to be disappointing, the thinner wing raising questions as to whether the new aircraft could attain a satisfactory ground performance. Various options were available to improve the situation. These included installing the Konus in-flight refuelling system, and using high-lift devices such as boundary layer blowing with wing flaps or improving the lift/drag ratio of the wing by using boundary layer suction. However, all of these solutions would either add weight to the structure or significantly erode the fuel economy of the engines (through their need to provide bleed air so that the boundary layer suction could operate).

The Tu-138-60 and Tu-138-100 had identical estimated flight performance figures (design data for this aircraft is given in the table). Prior to the tunnel testing suggesting otherwise, it had been expected that their maximum range would be in the region of 2,400km (1,492 miles) although the interception range was 1,800km (1,119 miles) when flying at subsonic speeds and 1,000km (622 miles) when flown supersonically (intercepts could be made at 1,600km or 994 miles when flying a mix of subsonic and supersonic speeds). Estimated patrol endurance was between four and four and a half hours but it was the radar and weapons that brought the difference between the two. The Tu-138-60 had a radar detection range of 90km to 100km (56 to 62 miles), a target acquisition range of 60km to 65km (37 to 40 miles) and a missile launch range of 32km (20 miles); the equivalent figures for the Tu-138-100 were all rather

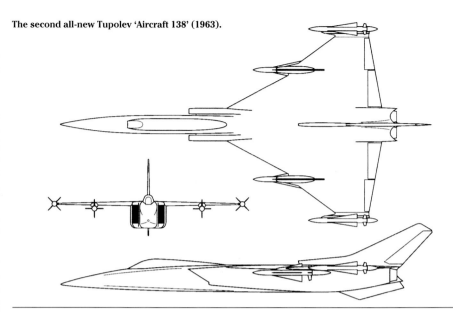

The second all-new Tupolev 'Aircraft 138' (1963).

higher – 130km to 200km (81 to 124 miles), 90km to 150km (56 to 93 miles) and 60km to 70km (37 to 44 miles).

Data is available only for the Tu-138-60 and -100, but two other quite different layouts were also studied during 1963. Both had wings mounted high on the fuselage, a single main fin and rudder and a single ventral fin, box-shaped side intakes and twin engines (believed to be VD-19s or the RD-36-41 derived from it) side-by-side in the rear fuselage. However, one had a canard and a relatively small cranked delta wing, the other no canard or tailplane but a larger delta wing with just a minimal reduction in sweep angle on the outer section. Each carried four air-to-air

missiles, which appear to have been versions of the K-80 (R-4) carried by the Tu-128. The canard type had two mounted at the wingtips and two more on pylons fitted to the tips of the canards, the pure delta carried its weapons on two mid-wing and two wingtip pylons.

None of the 'Aircraft 138' designs were selected for construction. By the middle of the 1960s the Tupolev OKB had decided to develop a new aircraft using a VG wing under the designation 'Aircraft 148' (which is described in Chapter Nine). The opening of design work on the '148' brought all further work on 'Aircraft 138' to a close. The R-60 missile was cancelled and the designation was used later for another weapon.

1960s Fighter Designs – Data / Estimated Data

Project	Span m (ft in)	Length m (ft in)	Gross Wing Area m² (ft²)	All-Up-Weight kg (lb)	Engine kN (lb)	Max Speed / Height km/h (mph) / m (ft)	Armament
Mikoyan 23-01 (flown)	7.72 (25 4)	16.6 (54 5.5) including probe	26.5 (284.9)	12,500 (27,557)	1 x R-27F-300 51.0 (11,465), 83.3 (18,745) reheat + 2 x RD-36-35 23.0 (5,180)	None available	4 x K-23 AAM, 1 x 23mm cannon
Mikoyan 23-11 (flown)	13.965 (45 10) at 16° sweep, 7.779 (25 6) at 72° sweep	16.71 (54 10) with pitot	37.27 (400.8) at 16° sweep, 34.16 (367.3) at 72° sweep	13,300 (29,321)	1 x R-27F-300 51.0 (11,465), 76.4 (17,195) reheat	2,240 (1,392) Mach 2.12 at 13,600 (44,619) clean, 2,025 (1,259) Mach 1.91 at 12,800 (41,995) with 4 AAMs	4 x K-23 AAM. (Planned 1 x 23mm cannon not installed in prototype but carried by production aircraft)
Sukhoi Su-17 (flown)	13.656 (44 9.5) (30° sweep), 9.64 (31 7.5) (63° sweep)	16.415 (53 10) with pitot	38.52 (414.2) (30° sweep), 35.85 (385.5) (63° sweep)	16,949 (37,366) with bombs and drop tanks	1 x AL-7F-1-250 66.6 (14,990), 94.0 (21,160) reheat	1,200 (746) at S/L, 2,150 (1,336) at height clean, 1,400 (870) at height with 4 bombs	Up to 3,000kg (6,610 lb) bombs plus stores, including AAMs on some marks
Mikoyan MiG-25P (flown)	14.015 (46 0)	21.42 (70 3.5)	61.4 (660.2)	36,720 (80,952) with AAM	2 x R15B-300 73.5 (16,535), 100.0 (22,510) reheat	1,300 (808) at S/L, 3,000 (1,865) Mach 2.83 at height	2 x R-40R + 2 x R-40T AAM
Tupolev 'Aircraft 138'	17.53 (57 6)	31.73 (104 1)	?	45,000 to 47,000 (99,206 to 103,616)	2 x VD-19 127.4 (28,660) reheat	2,100 to 2,400 (1,305 to 1,492) with missiles at height	4 x K-60 or K-100 AAM

Chapter Nine

Into the 1970s

The 1970s and 1980s witnessed a big change in the development of Soviet fighter aircraft. Until now the nation's economy had been able to support substantial numbers of competitive design procurements with multiple prototypes, but the ever-increasing cost of military aircraft brought this policy to a end. The remarkable quantity of different fighters flown by the Soviet Union has been noted several times already, but from the 1970s the number of new aircraft dropped off. In fact the cost of producing so many types had been hugely expensive and the Soviet Union's massive financing of defence equipment would eventually help to bring about the end of this communist superpower. However, those types that were flown during this later period were far more complicated aeroplanes because they had to introduce advanced and sophisticated avionics and better quality materials. Needless to say the variety of designs actually proposed was still as varied as ever.

Mikoyan MiG-31 Background

The MiG-25 was not only a very fine and successful design in its own right but it also served as a stepping stone towards the creation of one of the world's finest heavy interceptors. The Soviet Union still wanted better air defences in its Polar regions but, in the late 1960s, the Tupolev Tu-128, MiG-25PD and Sukhoi Su-15TM interceptors that equipped the PVO in this area were hampered by a relatively limited range. Their weapons and control systems were also becoming outdated, so the need for a new interceptor would very soon become quite urgent.

As early as 1965 Mikoyan began looking at ideas for a replacement which was intended to have the Smerch-100 (Tornado-100) fire control radar and the K-100 long-range AAM. Concurrently, the Yakovlev and Tupolev Design Bureaux also worked on long-range heavy interceptor projects, the latter with the 'Aircraft 148' described below. At the same

Mikoyan MiG-29 first prototype.

time the Volkov Bureau launched a large-scale research effort into designing a new 'look-down/shoot-down' radar for fighters that would give them the capability to destroy targets flying below their own flight level; previously ground clutter had complicated the target tracking and lock-on process by giving false radar returns.

Such was the progress in systems development that by the end of the 1960s the Smerch-100 no longer met current requirements and so the development of a new intercept system, called S-155, was launched by a joint Communist Party Central Committee and Council of Ministers directive dated 24th May 1968. First and foremost this was intended to counter the threats posed by new strike and reconnaissance aircraft such as the American General Dynamics FB-111A fighter-bomber and Lockheed SR-71 Blackbird Mach 3

One of the earliest configurations for Mikoyan's Ye-155MP project, showing the variable-geometry wings in their maximum sweep positions (c1968).

This very similar design shows the wings at minimum sweep but other differences include twin folding ventral fins (shown here in the extended position), a brake parachute housing in the base of the fin and a new colour scheme.

Another version of the VG Ye-155MP project had side-by-side cockpit seats and a single pair of wheels on the main landing gear. Note also the missiles carried beneath the wing leading edge extensions.

reconnaissance aircraft. The same directive ordered Mikoyan to create an advanced version of the MiG-25 designated Ye-155M. Three different variants, an interceptor, a tactical strike aircraft and a reconnaissance type, were planned and, initially, the Ye-155M was seen purely as an upgraded MiG-25; the resulting aircraft, however, was an all-new design.

To begin with three alternative layouts were examined as potential configurations for the interceptor, which was known as the Ye-155MP. These differed mainly in their wing and vertical tail design, the fuselage and the old MiG-25's characteristic lateral air intakes remaining virtually unchanged. Version A had a three-spar trapezoidal wing with small leading edge root extensions and B featured a variable-geometry swing-wing. However, Version C was the most unconventional because it embraced a tailless-delta layout with an ogival wing of increased area that resembled the shape used by the MiG-21I 'Analog'. The MiG-21I was a version of the MiG-21 fitted with an ogive wing to serve as a scale-model technology demonstrator for the Tupolev Tu-144 supersonic transport and it flew in 1968. Gleb Lozino-Lozinskiy was appointed Ye-155MP chief project engineer.

The tactical strike and reconnaissance version was called the Ye-155MF and the pure reconnaissance variant was the Ye-155MR: they differed mainly in armament and equipment. As we shall see the VG version was the most favoured at the start because it offered the required range and endurance; however, as the design work progressed the swing-wings and some of the other radical innovations were rejected. Gradually the new design moved away from the MiG-25 until all they had in common was an approximately similar layout plus similar dimensions.

The control system designed for the aircraft and its weapons was called Zaslon (Barrier) and required a huge effort to bring it to fruition. The result, however, was very impressive in that it enabled the interceptor

Beneath view of the second and third 'versions' of the Mikoyan Ye-155MP showing the dissimilar arrangements for missile carriage. Note the different main undercarriages, the same nosewheel but in different positions, single ventral fin on the later model (which is folded for landing) and the alternative size of the horizontal tailplane. The larger (dark) missiles are Vympel K-33s, the slimmer white weapons K-100s.

to attack several targets at once. For the first time in Soviet design practice the organisation designing the fire control radar held overall responsibility for the future interceptor's entire weapons system, because the various items of equipment needed to be integrated together to allow the aircraft to operate with maximum effectiveness. In 1969 Zaslon's designers took the extremely daring decision to make the radar antenna non-moving and to use a radar beam that would be scanned electronically – this was a first because no other fighter had used such equipment before. The maximum tracking range for a medium bomber-sized target was to be of the order of 120km (75 miles), and for a fighter-type 90km (56 miles) when head-on to a target and 70km (43½ miles) when attacking from the rear.

The interceptor's new missile was called the K-33 and was developed by the Vympel OKB (formerly the Bisnovat Design Bureau). It was a long-range weapon that used semi-active radar homing and had folding fins to allow carriage in tandem pairs semi-recessed beneath the fuselage; this cut the aircraft's overall drag by a considerable margin. The maximum effective 'kill' range was 120-130km (75 to 81 miles) and targets could be attacked at altitudes between 50m and 28,000m (164ft to 91,864ft) when travelling at speeds of up to 3,700km/h (2,300mph). There were also two pylons under the wings to carry four short-range AAMs on double launchers, two medium-range AAMs or two drop tanks. Thus the Soviet Air Defence Force acquired the potential capability to repel massive enemy air raids including those carried out at low altitude. The interceptor itself would possess a much greater range and endurance, the specified figures being 700km (435 miles) range when cruising at 2,500km/h (1,554mph) or Mach 2.35 and 1,200km (746 miles) when flying at subsonic speed.

Within the overall project a number of alternative design proposals were considered and some of these progressed to the full-scale development stage. Designs with a fixed trapezoid wing, of which there were quite a number, were given the in-house code 'Izdeliye 518', although the Ye-155MP designation also embraced much of this work.

Mikoyan Ye-155MP

This interceptor study was really a cross between the MiG-23 and the MiG-25 and had two crew seated in tandem under a canopy that was very reminiscent of the American McDonnell Douglas F-4 Phantom II. The design had air intakes similar to the MiG-25's (with horizontal airflow control ramps) but these were more aerodynamically refined. The VG wings were shoulder-mounted and there was a single fin and rudder assembly with a prominent fillet quite similar to the MiG-23's. To ensure there was sufficient directional stability two large ventral fins were provided, which folded away when the landing gear was extended.

This design was to be powered by two Solov'yov D-30F afterburning turbofans. Its armament consisted of three or four K-33 long-range air-to-air missiles semi-recessed in the lower fuselage plus additional short-range AAMs carried on pylons under the fixed wing gloves. There was a tricycle landing gear but the main units were unusual in having four-wheel bogies with small wheels to reduce runway loading, a feature that would enable the aeroplane to operate from dirt or ice strips; the nose unit also had twin wheels. The swing wings not only improved the project's field performance but, in certain flight conditions, they would also increase the fighter's on-station loiter time. However, the sweep-change mechanism increased the empty weight and the airframe's structural complexity; in addition, unlike the MiG-23, this aircraft was not intended for dog fighting where swing wings might have conferred an advantage. Hence the variable geometry Ye-155MP was abandoned.

Mikoyan Ye-158

Little is known about this project for a tailless ogival delta wing interceptor, but it did not survive for very long. It was a twin-engined two-seat design and it is possible that this was Version 'C' of the original studies made under the Ye-155MP interceptor designation, or at least a derivative of it.

Mikoyan 518-21

A new fixed-wing design called 'Izdeliye 518-21' was proposed in 1968 and, in keeping with the official directive issued on 24th May that year, this interceptor was to commence its state acceptance trials during the fourth quarter of 1971. However, further research showed that the 518-21 layout would be overweight and short on rate of climb and service ceiling, so it was abandoned and a redesign followed. No dimensional, weight or estimated performance data is available for any of these designs.

Mikoyan 518-22

Work on another interceptor project under the Ye-155MP banner, designated 'Izdeliye 518-22', commenced in 1969. Two years later the choice of fixed trapezoidal wings, tandem seating for the crew and semi-recessed K-33 missiles had matured into its final form. The 518-22 featured prominent fold-down leading-edge root extensions plus leading-edge flaps and its powerplant consisted of two D-30F-6 afterburning turbofans. These were much more fuel-efficient than the MiG-25's R15B-300s, especially in subsonic flight, and they were used later by the Ye-155MP prototypes. Other changes from the MiG-25P included a new undercarriage with staggered main wheels, a three-spar wing, more internal fuel, the Zaslon radar and K-33 missiles.

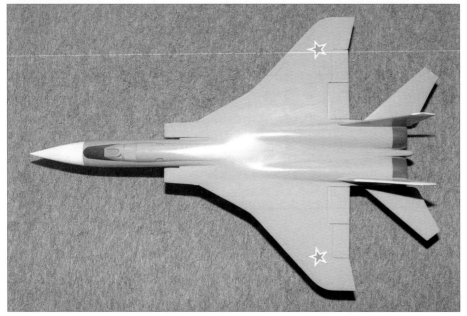

Top left and above: **Mikoyan's 518-22 design of 1969 had large leading edge root extensions which were only to be deployed during low-speed flight. For high-speed operations they folded downwards flush with the intake sides.**

Left and below left: **Mikoyan 518-55 design with MiG-29-style wing (c1970).**

Mikoyan 518-31

Another twin-engined two-seat interceptor project for which no further information is available. Many of the designs leading to the MiG-31 are in fact still classified at the time of writing.

Mikoyan 518-55

The '*Izdeliye* 518-55' had different wings that were rather like those of the MiG-29 fighter described shortly. Four recessed K-33 air-to-air missiles were carried under the fuselage, the main landing gear units had twin-wheel bogies with the wheels in line, the wheelbase itself was quite short, and the trapezoidal wings had large leading edge extensions and a kinked trailing edge. Once again there was a single ventral fin that remained folded for take-off and landing. This arrangement also remained a paper project.

Mikoyan MiG-31

Once the fixed trapezoid wing and twin fin layout had been selected for the Ye-155MP, full-scale design work could get going. This began in 1972 and, for the first time in Soviet fighter design practice, it was decided to equip the aircraft with afterburning turbofans. This move would ensure that the range requirements would be met and the choice was the D30F-6 developed by the Solov'yov Design Bureau. New main landing gears with twin-wheel bogies were also fitted that would allow the fighter to operate from unpaved airfields. For close-in combat a 23mm Gryazev/Shipoonov GSh-6-23 six-barrel Gatling cannon was mounted on the side of one of the air intake trunks, just aft of the main landing gear. Construction of the first Ye-155MP experimental prototype was completed in 1975 and the aircraft made its maiden flight on 16th September. The second machine flew in April 1976 and this had a full set of equipment, including the Zaslon radar, and also featured small under-fuselage fins.

Production aircraft entered service in 1983 as the MiG-31 and the ASCC named the type *Foxhound*. There were problems getting it into operation, but in service it has proved to be a truly impressive machine with excellent weapons and a phenomenal performance. Had it not been for the end of the Cold War, the MiG-31 would have presented an

This display model is also designated Ye-155MP and was the penultimate project in this series leading to the MiG-31. It has the 518-22's folding LERX (here retracted for high-speed flight) but a revised rear fuselage.

Plan view of the final Ye-155MP with the LERX raised for low-speed flight.

The first prototype Mikoyan Ye-155MP/MiG-31.

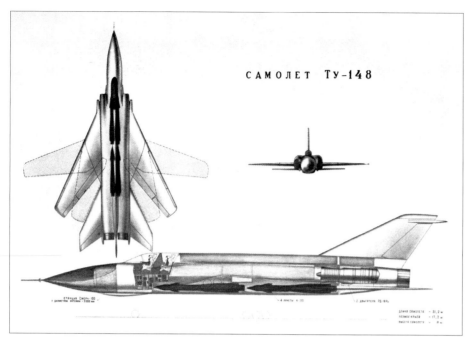

САМОЛЕТ ТУ-148

Tupolev Tu-148-100 project (1965). Note the three wing sweep positions and the number of undercarriage wheels, plus the large weapon bay shown here with four K-100 air-to-air missiles aboard.

immensely difficult opponent for the West's bombers and attack aircraft. The basic interceptor can operate over a wide speed and height band and has a top speed at altitude of 3,000km/h (1,865mph) and service ceiling of 20,600m (67,585ft); the limiting Mach number is 2.83. At sea level the maximum speed is about 1,500km/h (932mph), a capability the MiG-31 shares with few other fighters (the Panavia Tornado F Mk 3 is one). The internal fuel load totals 15,500kg (34,171 lb) and the maximum range with four R-33s plus external tanks approaches 3,000km (1,865 miles); range when cruising at Mach 2.35 is 1,400km (870 miles).

An improved version called the MiG-31M (*Foxhound-B*) began its development in the mid-1980s and introduced many changes, including the carriage of very long-range R-37 missiles plus R-77 AAMs. Seven examples were built, the first flying on 21st December 1985, but a lack of funding has prevented any orders being placed for the Russian Air Force.

Parallel Developments

Sukhoi Studies?
There have been rumours that Sukhoi studied an interceptor development of the Su-24 strike aircraft (described in the volume on bombers) as a rival to Mikoyan's work. It is entirely conceivable that Sukhoi would have undertaken research against the heavy interceptor requirement, but to date no concrete

information has been made available to confirm this.

Tupolev 'Aircraft 148'
Competition to the Ye-155MP did arrive in the form of the Tupolev 'Aircraft 148'. Work on updating the Tu-128 interceptor (Chapter Five), and the dead-end reached with the Tu-138 project (Chapter Eight), forced Tupolev to look in a different direction for a successor. As this work began almost at the same time as the 'Aircraft 145' swing-wing bomber, it seemed logical that the same type

of flying surface might offer a solution to the interceptor problem. Project work on the '148' began in 1965 and by the autumn Tupolev had come up with a general arrangement for the new design, together with its role and potential capability.

To go with the VG wing were two VD-19R2 turbojets (a development of the VD-19 afterburning engine) that offered an estimated potential top speed with the wings at or near maximum sweep of 1,400km/h (870mph) at between 50m to 100m (164ft to 328ft) altitude, and 2,500km/h (1,554mph) between 16,000m and 18,000m (52,493ft and 59,055ft). The practical operational range when flying at 2,500km/h was 2,500km (1,554 miles), but when flying at a height between 50m and 500m (164ft and 1,640ft) and a speed of 1,400km/h the operational range would be 570km (354 miles); at 1,000km/h (622mph) this was increased to 1,850km (1,150 miles). Finally, with the wing-sweep angle set for subsonic cruise flight a range of 4,800km (2,983 miles) was thought possible. Dependent on the interceptor's speed the range could be increased by another 30% to 40% through the use of in-flight refuelling.

The 148's estimated take-off weight was rather higher than the Tu-128, yet its take-off and landing performance would be signifi-

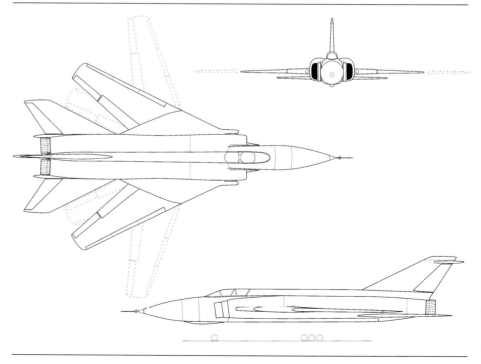

Line drawing of the Tu-148-100 project.

Soviet Secret Projects: Fighters Since 1945

cantly better. The take-off run would be 800m (2,625ft) and the aircraft was expected to be capable of using third-class and unpaved runways. This compared well against the Tu-128 which needed at least 1,350m (4,429ft) of runway to get airborne and at worst could only operate from a second-class airfield. In addition to its long-range intercept duties, the '148' was also designed to detect and disrupt the enemy's air traffic well behind its front line, to employ air-to-surface missiles against ground targets, undertake high and low-level reconnaissance, operate as a tactical bomber armed with conventional and nuclear bombs, and finally it could serve as a close-support aircraft. In fact 'Aircraft 148' was a full multi-role type that could, had it been built, have offered competition to the Soviet Union's new strike aircraft projects like the Sukhoi Su-24.

A key aspect of the '148's design was to be the Smerch-100 combined radar and missile guidance system. Like the Tu-128, the '148' was a large aircraft which allowing a large-diameter scanner to be installed – the Smerch-100 set intended for the '148' had a dish diameter of 2.0m (6ft 6⅜in). In the first stage of development it was planned to use K-100 heat-seeking air-to-air missiles armed with a selection of different warheads and capable of a range of up to 80km (50 miles). Once Smerch-100 was operational some longer range missiles could then be introduced so that targets could be intercepted and destroyed when flying at altitudes from 50m up to 35,000m (164ft and 114,830ft) and at speeds between 500km/h and 4,500km/h (311mph and 2,797mph). The onboard equipment included a trajectory control system, flight and navigation systems, an automatic data processing system for different types of target, and the reception and relay of data to other aircraft.

All the weapons and much of the avionics of the '148' were to be accommodated in a very large modular-style fuselage bay that was designed for a rapid switch from one operational role to another. The various armaments comprised four of the K-100 missiles, one Kh-22 cruise missile, two Kh-28 anti-radar missiles or four Kh-100P missiles, two tactical nuclear bombs, unguided missile launchers or cannon. Other equipment included Bulat (Sword), Sablya (Sabre) or Virazh (Turn) electronic intelligence gathering (Elint) gear, or various cameras or additional fuel tanks. The full weapon system would be known as the Tu-148-100 and, act-

ing as a long-range patrol interceptor, the aircraft would have an intercept range when flying subsonically of 2,150km (1,336 miles). For a 2,500km/h (1,554mph) flight speed this figure dropped to 1,000km (622 miles) and flying a mix of subsonic and supersonic speeds reduced it to 1,700km (1,057 miles). The '148' was expected to be able to patrol for up to two hours at a distance of 1,300km (808 miles) away from its base and four hours when 500km (311 miles) away. Thus the aircraft would be able cover the USSR's northern and eastern regions with comparatively little support.

If the project was successful the PVO would get its hands on a highly effective air-defence system, but the Achilles heel was the sophisticated avionics that would be needed to achieve this objective. Such was the state of the Soviet electronics industry in the 1960s that the Tu-148-100 system would take at least a decade to bring to fruition, an estimate later confirmed by the time that was needed to create the less complex but more effective

Zaslon system for the MiG-31 (Zaslon was not ready until the 1970s). In addition the concept of a multi-role aeroplane did not fit in with the outlook of the VVS who usually preferred to operate different types of specialist aircraft, so the Tu-148-100 project failed to interest the service and did not proceed any further. Work on it was abandoned in early 1965.

However, 'Aircraft 148' was revived in the second half of the 1960s after studies had begun with the new Zaslon weapon system combined with K-33 missiles. These offered a detection range of 110km to 115km (68 to 71 miles) and a missile launch range of 80km to 90km (50 to 56 miles). In its general design and function the system resembled the Smerch-100 but its capabilities were more modest and realistic. The idea of a multi-role aircraft for the VVS was abandoned and instead efforts were now concentrated on an up-to-date but realistic long-range interceptor which fell within the capability of the USSR's industry to produce. The '148' was thus no longer seen as a multi-role type but instead as

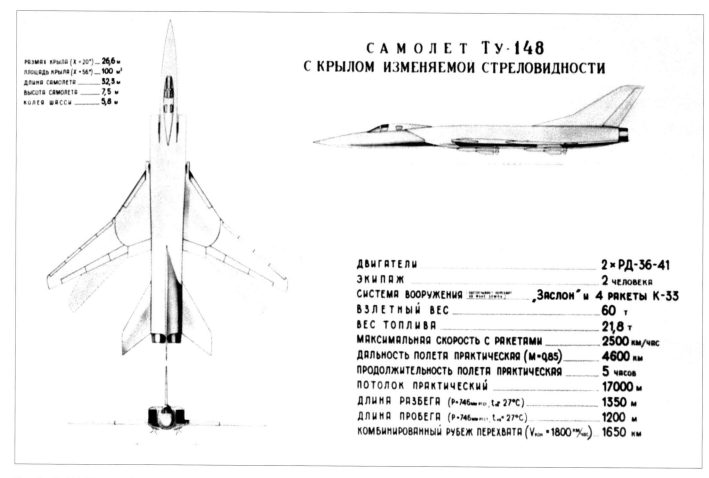

САМОЛЕТ Ту-148
С КРЫЛОМ ИЗМЕНЯЕМОЙ СТРЕЛОВИДНОСТИ

РАЗМАХ КРЫЛА (X = 20°) — 26,6 м
ПЛОЩАДЬ КРЫЛА (X = 56°) — 100 м²
ДЛИНА САМОЛЕТА — 32,5 м
ВЫСОТА САМОЛЕТА — 7,5 м
КОЛЕЯ ШАССИ — 5,8 м

ДВИГАТЕЛИ	2 × РД-36-41
ЭКИПАЖ	2 человека
СИСТЕМА ВООРУЖЕНИЯ	„Заслон" и 4 ракеты К-33
ВЗЛЕТНЫЙ ВЕС	60 т
ВЕС ТОПЛИВА	21,8 т
МАКСИМАЛЬНАЯ СКОРОСТЬ С РАКЕТАМИ	2500 км/час
ДАЛЬНОСТЬ ПОЛЕТА ПРАКТИЧЕСКАЯ (М=0,85)	4600 км
ПРОДОЛЖИТЕЛЬНОСТЬ ПОЛЕТА ПРАКТИЧЕСКАЯ	5 часов
ПОТОЛОК ПРАКТИЧЕСКИЙ	17000 м
ДЛИНА РАЗБЕГА (Р-746.... t.= 27°C)	1350 м
ДЛИНА ПРОБЕГА (Р-746.... t.= 27°C)	1200 м
КОМБИНИРОВАННЫЙ РУБЕЖ ПЕРЕХВАТА (V.... =1800 ᵐ/....)	1650 км

Tupolev Tu-148-33 project (late 1960s). The drawing shows four K-33 missiles carried externally.

a further upgrade of the Tu-128 fitted with VG wings.

Compared with the Tu-128S-4 system, the new Tu-148-33 would provide improved operational effectiveness in several ways:

i. The ability to intercept and shoot down targets flying at low level and at altitudes between 500m and 8,000m (1,640ft and 26,247ft).

ii. The capability to deal with an increase in the speed of head-on targets from 2,000km/h (1,243mph) to 3,500km/h (2,175mph) and behind from 1,250km/h (777mph) to 2,400km/h (1,492mph), together with an increase in their height from 21,000m (68,898ft) to 28,000m (91,864ft).

iii. The ability to intercept small targets such as missiles.

iv. It would be able to engage two targets simultaneously and to operate as one of a group of aircraft.

v. There was improved resistance to enemy jamming.

vi. The aircraft had greater range and longer endurance, improved take-off and landing performance and improved acceleration.

The Tu-128's AL-7F-2 engines would be replaced by more powerful Rybinsk RD-36-41 units and there would be alterations to the fuselage, air intakes and ducting. The VG wing would have double-slotted internal flaps, the undercarriage would be strengthened and it would have new wheels. Aileron-spoiler control and automatic stability and flight control systems would also be introduced and the Tu-148-33 would be able to use the same airfields as the Tu-128. The aircraft had a crew of two, the internal fuel load totalled 21,800kg (48,060lb) and the estimated service ceiling was 17,000m (55,774ft). Range at subsonic speeds was 4,600km (2,859 miles), falling to 1,650km (1,025 miles) when flying a combination of subsonic and supersonic speeds. It was this aircraft that presented the real competition to the future MiG-31 programme.

The Tu-148-33 project was submitted to PVO Command and received the support of its Commander-in-Chief General A Kadomtsev. Tupolev therefore continued with the aircraft's design and its weapon system and duly prepared a full-scale mock-up that was inspected on several occasions by PVO and VVS representatives. Then in May 1968 Kadomtsev was killed in a crash involving

one of the first Ye-155P (MiG-25) fighters and his replacement opted instead to update the MiG-25P with the Zaslon system. As a result serious work began on the Ye-155MP, which of course became the prototype for the future MiG-31. However, PVO Command did not reject the Tu-148-33 project immediately and there were rumours that Tupolev's design would still be the preferred choice, but in truth this was the end. New requirements were introduced, including much improved manoeuvrability at low altitudes and increased rates of climb and service ceiling, and these were at odds with the original '148' concept. The changes that would be needed to accommodate the new limits would require the removal of the VG wing.

The reason for delaying the formal rejection was that Andrei Tupolev and his closest aides tried to change PVO Command's opinion and to obtain a government resolution for their system, or at least get further funding to allow work to continue. Tupolev's word carried great weight in aviation circles and so he could approach officials at the highest level. He tried to generate more interest in the '148' by offering a naval version suited for carrier operations, but Tupolev's efforts were in vain. In fact the new improved official require-

Early Mikoyan MiG-29 twin-engine project with a single fin and rudder (c1971). This project, or a version of it, appears to have been Mikoyan's first submission to the PFI contest.

ments had, in part, been raised to prevent a row with the USSR's most senior and respected aircraft designer. All work on 'Aircraft 148' was finally terminated at the start of the 1970s and the Tupolev Design Bureau would never again return to the design of large long-range interceptor aircraft for the PVO, apart that is from some technical proposals made for this type of aeroplane that were based on the Tu-144D supersonic airliner and Tu-22M bomber. The resulting long-range fighter-interceptor 'raider' concepts were called DP-1 and DP-2 respectively, but neither progressed beyond an initial study.

MiG-29 and Su-27 Background

In about 1969 all three of the Soviet Union's Fighter Design Bureaux began to look at ideas for a fourth-generation fighter. Part of the work involved analysing the operational experience of current types in various regional conflicts so that the design teams might be able to give their new aircraft enhanced capabilities. Much of this effort was stimulated by the appearance of the American F-X fighter programme, which brought forth the McDonnell-Douglas F-15. The Soviets monitored the progress of this programme very closely because their response would have to be more than a match for the American aircraft, an immensely difficult objective to achieve.

In 1971 TsNII-30, a division of the Soviet Ministry of Defence, issued the first General Operational Requirement for a fourth-generation fighter, which tentatively was designated PFI (*perspektivnyy frontovoy istrebeetel* or advanced tactical fighter). The aircraft's primary tasks would be to deal with enemy fighters in close-in combat using new short-range AAMs or an internal gun, or to intercept aerial targets at more distant ranges and destroy them with new medium-range AAMs; the latter would be achieved either by using the aircraft's own 'look-down/shoot-down' radar or with guidance from ground control stations. The aircraft would also undertake other duties, such as ground attack, and compared to previous types it would be more agile and manoeuvrable and carry a set of all-new avionics. Maximum speed had to be at least 1,400km/h (870mph) at sea level and 2,500km/h (1,554mph) at 11,000m (36,089ft) and the maximum

Mach number was to fall between 2.35 and 2.5. Sea level rate of climb had to be at least 300m/sec (984ft/sec), service ceiling 21,000m (68,898ft), maximum range without drop tanks 1,000km (622 miles) at sea level and 2,500km (1,554 miles) at height, and the take-off thrust-to-weight ratio had to be between 1.1 and 1.2.

In 1972, after having revised the Operational Requirements, the VVS issued a request for proposals for new fighters. Mikoyan submitted two versions of its MiG-29 project, Sukhoi's entries were the T10-1 and T10-2 projects and Yakovlev entered the Yak-45I light lighter and the

Yak-47 heavy fighter. These designs and their background are described below and the outcome was to be a mix of light and heavy fighters for the Soviet Air Force. The seeds for this result came from some references made in 1971 by the Government's research establishments, which stated that the fighter fleet planned for the 1980s should be based on two types, one heavy and one light. This move followed American practice where it was becoming clear that the US Air Force's heavy F-15 Eagle would be complemented by a new lightweight fighter in a competition eventually won by the General Dynamics F-16 Fighting Falcon.

Heavyweight MiG-29 design with four Soviet K-25 'Sparrow' missiles on tandem pylons fixed to the corners of the lower fuselage underneath the wings. The design features leading edge slats plus twin fins and twin ventral fins mounted on short twin 'booms' (1972).

Mikoyan PFI Studies

Work on Mikoyan's fourth-generation fighter got under way in 1970 and the aircraft was designated MiG-29 at a very early stage. Gleb Ye Lozino-Lozinskiy was given overall control of the project and both TsAGI and GosNIIAS became heavily involved. The engineers looked both at conventional designs and also the so-called 'integral' layout, where the wings and the fuselage were blended together into a single lifting body. Much thought was also given to the powerplant. A nose intake was immediately ruled out because the inlet ducting would occupy a disproportionately large portion of the internal volume; consequently two lateral intakes would have to be used. An initial design powered by a single large turbojet (and tunnel tested at TsAGI) was quickly rejected and, basing its decision on operational experience with the best third-generation single-engined fighters, Mikoyan soon chose a twin-engined design.

One early configuration had shoulder-mounted trapezoidal wings, a low-mounted horizontal tail and a single fin and rudder. The wings featured leading-edge root extensions and full-span leading-edge slats and the project also had sharply raked two-dimensional air intakes with horizontal airflow control ramps that were very reminiscent of the MiG-25. There were four underwing hardpoints for short-range 'dogfight' missiles and two more were tucked onto the corners of the intake boxes to carry medium-range AAMs. The fighter's thrust-to-weight ratio was 1.12, estimated service ceiling 21,500m (70,538ft), maximum rate of climb at sea level 290m/sec (951ft/sec) and range without drop tanks 820km (510 miles) at sea level and 2,000km (1,243 miles) at height.

A second design also had MiG-25 box-style air intakes and had twin wheels on its the nose gear but, in contrast to the MiG-31's staggered format, the main units featured tandem twin-wheel bogies with the wheels located in line (rather like the Swedish SAAB 37 Viggen). This aircraft was to be armed with four medium-range K-25 missiles that were based

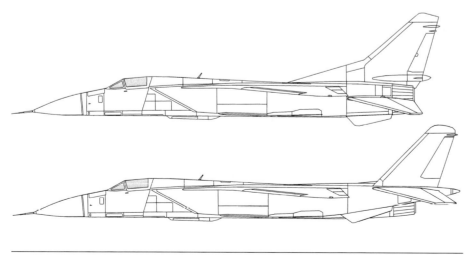

Drawings of two MiG-29 heavy fighter layouts. The upper project dates from 1971 and has a single fin, the lower is the twin-boom project from 1972. The forward fuselages appear to be identical.

Soviet Secret Projects: Fighters Since 1945

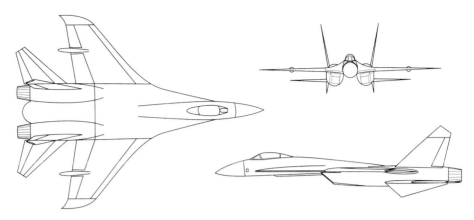

Sukhoi's first T10-1 configuration (1970). Note the small wing fairings between the ailerons and flaps to house the bicycle undercarriage's outrigger legs.

Model of Sukhoi's earliest T-10 layout.

on the American Raytheon AIM-7E Sparrow, a handful of which had been captured in Vietnam and shipped to the USSR for examination. However, work on this missile was eventually discontinued.

A later study for a lightweight fighter also showed an integral layout with the fuselage, wings and engine nacelles all blended together. Compared to the 'heavy' projects above, this had a normal gross weight of around 12,800kg (28,219 lb), which made it not only lighter than other Mikoyan designs but also lighter than Sukhoi's Su-27/T-10 entry for the PFI competition (below), which had a maximum take-off weight of 21,000kg (46,296 lb). This project was actually shorter than the eventual MiG-29 and had a wing area of only 25m² (268.8ft²). One advantage of the integral layout was that the prominent LERX provided a convenient location for the internal gun – finding a suitable place for it aboard 'conventional' layouts had actually turned into quite a problem. Mikoyan's integral design had its opponents and some engineers believed that a conventional fighter would be easier to build and would offer a smaller maximum cross section.

One of the heavy MiG-29 studies, believed to be the design with the single fin, was submitted to the PFI competition in 1972. However, two months after the first discussion meeting had been held during that year, Mikoyan proposed this lightweight project as an official second entry.

Sukhoi PFI Studies

The first general arrangement sketches for what became Sukhoi's efforts to the new requirements were completed as early as the summer of 1969. However, to begin with Pavel Sukhoi was not keen to take part in the competition because he felt that current Soviet avionics were too heavy and they would prevent the development of a fighter that could match the requirements. In addition his Design Bureau was full of work on upgrades to the Su-15 and Su-17 fighters, refining the Su-24 strike aircraft and producing developments of the T-4 bomber.

Some months later Sukhoi was forced to start work on a fighter design and the resulting project, unveiled in February 1970, was another integral layout with the wings and fuselage blended into a single body; it also

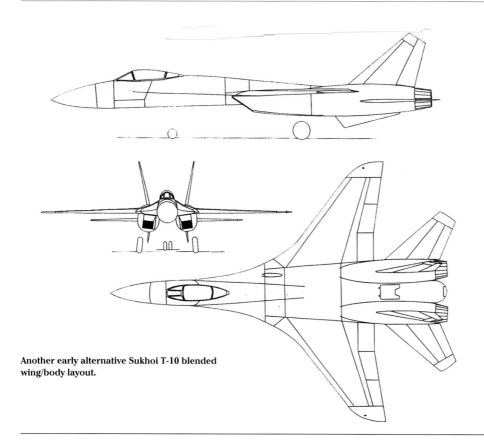

had twin tails and low-mounted stabilisers. The engines were housed in underslung nacelles spaced well apart and, as a whole, the integral arrangement was inspired by Sukhoi's unbuilt T-4MS bomber project (see the companion volume on bombers). Special ogival wings were fitted which had a complex curvature at the centre section, but the leading edge sweep angle changed smoothly along the whole span. A special bicycle undercarriage was fitted because problems had been experienced in trying to work in a tricycle arrangement. The design was the work of Vladimir Antonov, Oleg Samoylovitch and Valery Nikolayenko.

In 1971 another team of engineers led by A M Polyakov and A I Andrianov designed an alternative layout that was based on more traditional lines. It showed a conventional non-blended fuselage of almost rectangular section with no dorsal spine, lateral box-shaped air intakes and shoulder-mounted wings; it also had outward-canted twin fins and low-mounted slab stabilisers with marked anhedral. A certain amount of conservatism was prevalent among many engineers and officials who felt that a more

Another early alternative Sukhoi T-10 blended wing/body layout.

Drawings showing the Sukhoi T10-1 at the 'conceptual design' stage in 1972.
Nikolai Gordyukov

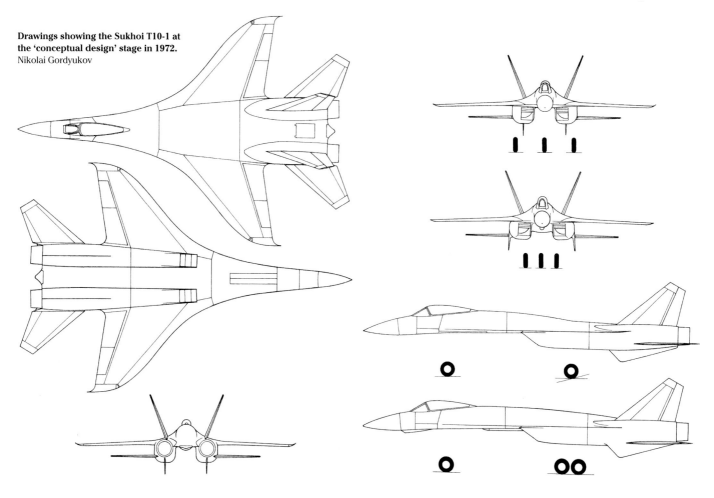

Soviet Secret Projects: Fighters Since 1945

conventional design would be superior and part of their argument was based on the fact that the F-15 did employ what could be described as a classic aerodynamic configuration. This second complementary project, however, did share the ogival wing but here it was not combined with the fuselage to form a single lifting body. Incidentally, it is worth noting that the North American company's contender to the F-X/F-15 requirement also had swept ogival wings with leading edge extensions.

The overall effort to design a new fighter received the in-house code T-10, with the blended wing/body Version A labelled T10-1 and the more conventional Version B T10-2 (these titles should not to be confused with the identical designations given by the manufacturer to the prototypes when they were first built). Officially the project became the Su-27. Jokes were circulating that the T-10 was a 'variable aerodynamic configuration aircraft' but Pavel Sukhoi gave his support to the blended wing/body design.

During 1972 both the T10-1 and T10-2 reached their 'conceptual design' stage ready to enter into competition with Mikoyan and Yakovlev. By now the 'integral' project's designers had worked in a tricycle undercarriage that in one form had single main wheels but in the primary submission twin tandem wheels; these retracted into wells in between the nacelles. The first design had its main gears spaced well apart, the latter closer together which, as the drawings show, affected the position of the ventral fins. Wing and fuselage were very smoothly blended as one and the engines were still widely spaced. The radar was housed in the nose, the fuel tanks were put into the mid-fuselage section, the fins were canted outwards and the horizontal all-moving tailplanes had slanted axes of rotation; there were also ventral fins on the sides of the engine nacelles. Six underwing hardpoints plus another under each intake box allowed carriage of up to eight air-to-air missiles, comprising six short-range K-60s which used infra-red homing plus two K-25 medium-range weapons with semi-active radar homing. Estimated thrust/weight ratio was 1.12, sea level rate of climb 345m/sec (1,132ft/sec), service ceiling 22,500m (73,819ft) and range without drop tanks

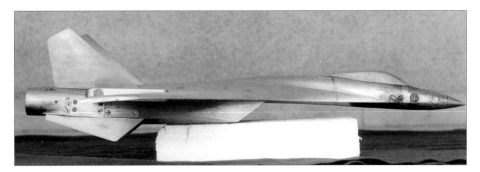

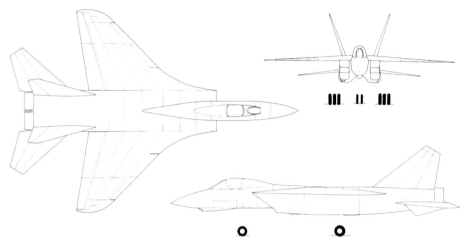

Wind tunnel model of the blended wing/body T10-1 'conceptual design'. Victor Drushlyatov

Sukhoi T10-2 'conventional' layout from the 'conceptual design' stage (1972). Nikolai Gordyukov

Model of the 'conventional' Sukhoi T10-2. John Hall

800km (497 miles) at sea level and 2,400km (1,492 miles) at height.

On the 'conventional' design the equipment and fuel were laid out in a roughly similar form to the blended version with the main gears (which had three small wheels side-by-side on each leg) folding away underneath the air ducts. The nose gear had two wheels and the twin-barrel 30mm cannon was mounted beneath the starboard intake duct ahead of the undercarriage well. There were some high-lift devices on the wings including two-segment flaps and deflectable slats, and the project had slightly smaller ventral fins. Again there were six underwing hardpoints plus two more beneath the middle fuselage to take AAMs, and the avionics suite was similar to the T10-1; this included a Sapfir-23MR (Sapphire) radar. Estimated performance data for this version is unavailable but models of both versions were tested in TsAGI's wind tunnels and it soon became clear that the integral layout was superior, mainly because it showed less drag during supersonic cruise. Consequently work on the T10-2 was brought to a close, but this decision was made after both T-10 projects had been submitted to the PFI competition.

Yakovlev PFI Studies

Yak-45: Leading up to the PFI competition the Yakovlev Design Bureau undertook a series of studies for a single-seat aircraft equipped with radar and missiles. The initial project bore the designation Yak-45 and was to be powered by two 78.4kN (17,635 lb) thrust Favorskiy 'Izdeliye 69' non-vectoring augmented turbofans, an engine derived from the R-28. Yak-45 also appears to have embraced a light attack aircraft design and it had large canard foreplanes with the power units underslung ahead of broad delta wings; as a result, Yakovlev expected the aircraft to possess outstanding manoeuvrability.

Yak-45I: The Yak-45I (some sources misidentify it as 'Yak-45M') was one of the studies made within the Yak-45 series. This was submitted for the PFI contest but, whereas Mikoyan and Sukhoi's primary contenders generally had a blended wing and body, the Yak-45I revealed a more conventional form. Unlike the original Yak-45, it had no canard foreplanes and was fitted with a horizontal tail of cropped-delta planform. A published drawing shows that it had broad compound delta wings with the engines, two 80.4kN

(18,080 lb) R53F-300 afterburning turbojets, mounted on the wings (rather than ahead of them). The nozzles protruded beyond the wing trailing edge and the air intakes featured protruding shock cones. The wings had approximately 40° of leading-edge sweep outboard of the engines and 75° inboard, the latter serving the same purpose as the leading-edge root extensions on the Mikoyan and Sukhoi proposals. On the wingtips were long slender fairings that were possibly intended to house outrigger wheels, which indicated that a bicycle undercarriage would have been fitted.

Yak-45I's thrust-to-weight ratio at take-off was 1.18, estimated service ceiling 21,500m (70,538ft), sea level rate of climb 340m/sec (1,115ft/sec) and range without drop tanks 1,000km (622 miles) at sea level and 2,500km (1,554 miles) at height. The 'I' in Yak-45I referred to *Istrebitel* or 'fighter' and reflected that it was only part of Yakovlev's proposed design programme. This series of designs was also intended to embrace strike aircraft, reconnaissance and vertical take-off types (the V/STOL study has been discussed briefly in Chapter Six) and all of these projects would have employed the same aerodynamics. The Yak-45I fighter was rejected by the PFI judges, after which work on the project was stopped.

Yak-47: Yakovlev's second submission to the PFI competition was this fighter that, in effect, was a scaled-up and heavier version of the Yak-45I with two 122.5kN (27,560 lb) R59F-300 afterburning turbojets. There were no protruding shock cones in the engine air intakes and the armament was unchanged from the Yak-45I, namely two K-25 and two K-60 air-to-air missiles. Tip nacelles again suggest that this project was another Yakovlev design to possess a bicycle undercarriage. The Yakovlev Bureau had decided to try and satisfy both the light and heavy elements of the requirement with, not so much the same aircraft but the same development effort (this was after the contest had been divided into light and heavy tactical fighter categories). The result was essentially the same design but in two different weight classes, with Yakovlev hoping to fill both requirements. Yak-47's take-off thrust-to-weight ratio was 1.1, estimated service ceiling 20,000m (65,617ft), sea level rate of climb 275m/sec (902ft/sec) and range without drop tanks 1,000km (622 miles) at sea level and 2,500km (1,554 miles) at height.

In the long run, however, Yakovlev's generally outdated proposals rendered them incapable of achieving the required agility and this helped to ensure that each version

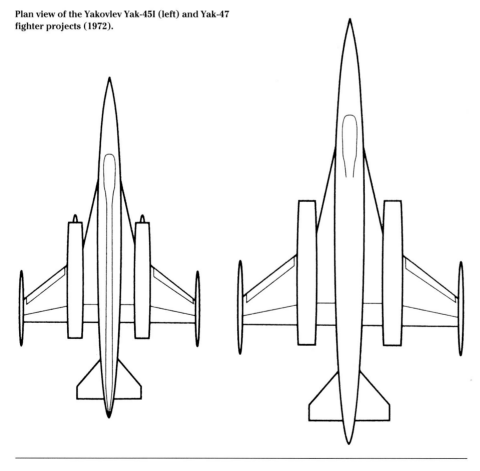

Plan view of the Yakovlev Yak-45I (left) and Yak-47 fighter projects (1972).

Artist's impressions of the Yak-45 contender to the PFI competition.

was turned down. Both designs were also expected to carry the smallest missile armament of all the contenders yet they were still heavier than their light and heavy competitors; work also ceased on the Yak-47 after Mikoyan and Sukhoi's designs had been selected.

–

A joint meeting convened between representatives of the Air Force together with the Scientific Technical Council of the Ministry of Aircraft Industry was held in 1972 and included representations and reports on these projects by their designers – Gleb Lozino-Lozinskiy (Mikoyan), Oleg Samoylovich (Sukhoi) and the veteran Designer General Aleksandr Yakovlev himself. At this stage Mikoyan's sole proposal was the 'heavy' MiG-29 but, when a second meeting was called two months later, the Bureau then presented its new 'lightweight' version as well. The two meetings brought the rejection of both Yakovlev designs, principally from concerns regarding their aerodynamics in an 'engine-out' situation – if one engine did fail it was felt that safe flight might be difficult to achieve. It was agreed that both the Mikoyan and Sukhoi proposals should proceed to the next stage of assessment.

At this point however, a new idea was suggested by Mikoyan, namely splitting the PFI requirements into two separate aircraft programmes. The heavy multi-role tactical fighter element could be filled by Sukhoi's T-10/Su-27 and its lightweight companion by the smaller MiG-29; this would ensure that both OKBs had new work while the two types could make use of common systems and equipment. In the past the USSR's Air Forces had operated several types of fighter at once, some of them just in a single role, but limitations with available funding would prevent such practice in the future. However, the need for fighters to undertake high-speed interceptions using ground control, or long-range escort missions, close-in dog fighting and various other duties, could not be filled by a single airframe. Mikoyan's proposal was accepted and the decision saved a great deal of time and effort that would have been needed to find a winner. At the end of 1972 official directives were ordered for the go-ahead of both the MiG-29 and Su-27, a move that left Mikoyan very happy (and more so when it received orders for the MiG-31 as well). Some studies were made to assess the optimal ratio of light to heavy fighters and this was found to be two thirds MiG-29 and one third Su-27.

Mikoyan MiG-29

Work on Mikoyan's fighter continued but proponents of the conventional layout continued to criticise the integral style of design. In early 1973 one of the Bureau's team leaders, A Nazarov with the backing of other OKB designers and engineers, reported that an integral layout gave a bigger surface area and hence created more drag than a conventional aircraft; his colleagues also claimed that the integral layout would not give sufficient structural strength. Yakov Seletskiy, the principal 'creator' of the integral format, together with Anatoliy Pavlov stood against the traditionalists and eventually the 'innovators' won their battle. It was found that in tight turns during a dogfight a conventional layout with lateral intakes showed a poor performance because the 'leeward' intake operated in extremely unfavourable conditions; moreover, the 'conventional' fighter could not attain angles of attack in excess of 20°.

However TsAGI, like the Design Bureau, had its share of traditionalists too who insisted that both the MiG-29 and T-10 should be produced along conventional lines. G S Büschgens, the director of TsAGI, urged Rostislav Belyakov (who had succeeded Artyom Mikoyan as the OKB's head) to abandon the integral design. Other influential TsAGI engineers did likewise and campaigned for a conventional layout. Belyakov was unwilling to speak openly on this issue because, as yet, he did not have enough political weight to make his comments tell. Belyakov did not participate directly in the development of the MiG-29. A A Choomachenko, who on 29th December 1972 was appointed MiG-29 project chief and Mikoyan OKB Deputy General Designer, proved a big help in confirming the choice of the integral layout over the conventional type. His team did a lot of work comparing the designs, analysing their strengths and weaknesses, and the results were then sent to TsAGI. These changed the minds the Institute's leaders and so, from now on, the integral MiG-29 design received favour from official circles as well.

Alongside the PFI competition and the studies that came from it, from the late 1960s four Soviet engine Design Bureaux, together with TsIAM (the Central Institute of Aero Engines), had been looking at new powerplants. The OKBs were Klimov (now led by Klimov's successor Sergey Izotov), the Soyuz (Union) plant in Moscow (Bureau led by Sergey Tumanskiy), Lyul'ka and finally the Perm' OKB led by Pavel Solovyov. All previous

This 'lightweight' Mikoyan MiG-29 study carries six missiles including four K-60s under the outer wings.

Soviet Secret Projects: Fighters Since 1945

Soviet fighter engines had been turbojets but both Lyul'ka and Solovyov produced designs for afterburning turbofans, the AL-31F and D-30F6 respectively; however, these were too big for a light fighter like the MiG-29 being rated in full reheat at 122.5kN (27,555 lb) and 151.9kN (34,170 lb) thrust respectively. The MiG-29 required a turbofan in the 80kN (18,000 lb) class and Izotov offered such an engine, the 81.3kN (18,300 lb) RD-33. Before long however, some competition appeared in the form of Tumanskiy's R67-300 turbofan rated at 73.5kN (16,535 lb) in reheat and so an unofficial contest was under way.

In 1973, after carefully studying the two powerplants, Mikoyan selected the RD-33 and the prototype engine first ran late in 1974. One potential problem was that the MiG-29's low intakes made them very vulnerable to foreign object damage but Mikoyan's designers found an unconventional and singularly effective solution that had not previously been used by any other fighter aircraft. On the ground the main air intakes were blanked off completely by perforated panels (foreign object protection doors), and the engines breathed through a series of blow-in doors on the upper sides of the LERX. During take-off rotation these dorsal blow-in doors were automatically closed while the protection doors swung open, which then allowed the engines to operate as normal.

The MiG-29 was to receive some very sophisticated avionics including three targeting systems – a pulse-Doppler fire control radar with 'look-down/shoot-down' capability, an infra-red search and track system and a helmet-mounted sight. However, the cost of the full standard aircraft was such that, for a period, a cheaper 'stop-gap' version, which was to use the MiG-23's avionics and armament, proceeded in parallel. The full aircraft with the original S-29 radar was known in-house as the '*Izdeliye* 9.11' while the latter became the 9.11A (or MiG-29A). Each version used identical airframes and powerplants and shared some systems, and their flight performance was near identical. Then on 19th January 1976 a directive was issued ordering the complete development of both the full standard MiG-29 and the Su-27 and this signalled the end of the MiG-29A; Mikoyan could now concentrate wholly on its MiG-29 as originally proposed.

On 15th July 1974 Belyakov approved an increase in wing area from 34m² (365.6ft²) to

Mikoyan MiG-29 study with six missiles, large LERX and a constant angle of sweep along nearly all of the wing leading edge. This design resembles the MiG-29 as built (mid-1970s).

The most extreme form of wing/fuselage blending was exploited on this late version of Mikoyan's lightweight MiG-29 fighter studies. Apart from this the model shows many similarities to the previous models, which show nicely the parameters that were examined in depth during this period of research (mid-1970s).

Drawing of the fully-blended MiG-29A variant.

38m² (408.6ft²) and the re-winged fighter received a new manufacturer's designation, 'Izdeliye 9.12'; the design was finally frozen in 1977. Unlike the Su-27, which was designed to operate well beyond the frontline, the MiG-29 would mostly fly over friendly territory. The air-to-air missiles developed for the new fighter were the medium-range Bisnovat K-27 and the short-range Bisnovat R-60M or K-73. Prototype 9.12 was completed in 1977 and first flew on 6th October and it showed a rate of climb of 330m/sec (1,083ft/sec) at 1,000m (3,281ft) and a service ceiling of 17,500m (57,415ft). Since then the MiG-29, codenamed *Fulcrum* in the West, has been a great success with home and overseas sales and several new versions. A naval variant for carrier operation called the MiG-29K, fitted with an arrester hook and other 'naval' items, first flew on 23rd June 1988. Other more advanced developments have since made their appearance and on 29th November 1997 Mikoyan flew the first MiG-29SMT prototype, which introduced an expanded upper rear fuselage spine to house more fuel together with on-going improvements to the avionics.

Sukhoi T-10 and Su-27

Returning to the Sukhoi side of the story, the T10-1's centre of gravity was located well aft, which made the aircraft statically unstable in pitch and thus enhanced its potential manoeuvrability. The aircraft was to become the first Soviet aeroplane to use (in planned production form) an automated fly-by-wire control system with no direct mechanical link between the stick, pedals and control surfaces; this feature contributed much to the success of the definitive Su-27 and the family of aeroplanes that stemmed from it. However, the T-10 project contained a lot of risk because many of its design features were new to the Soviet aircraft industry, yet Sukhoi was well aware that only an unconventional airframe could offset the traditional shortcomings of Soviet avionics, which were bulkier and heavier than their Western counterparts. The big question was would the ogival wing give the high manoeuvrability that Sukhoi wanted?

Detail design of the T-10 got going in 1973 and late in the year Naum Chernyakov was

Soviet Secret Projects: Fighters Since 1945

Model of the 1973 'packaged' intake T10-6.

The T10-6A had axis-symmetric (round) intakes beneath the wing roots, which are not visible in this view (1973).

This Sukhoi T-10 pre-project design has box intakes and is much closer to the aircraft as eventually built. It is believed to date from 1975.

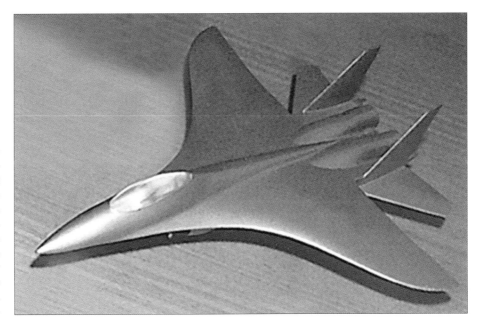

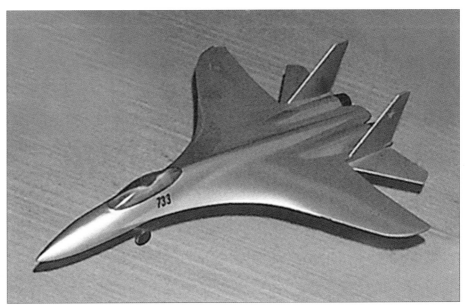

made chief designer. Between 1973 and 1975 a series of alternative designs was forthcoming, all of which featured ogival wings but introduced different intake arrangements and other changes. A problem that had to be dealt with first were the underslung engines with low-mounted intakes which were vulnerable to foreign object damage. To provide adequate ground clearance the landing gear had to be longer and heavier than on equivalent fighters with conventional fuselages. Therefore a revised project, designated T10-3, was drawn and this had its main gear units retracting into the bottoms of the engine nacelles along the lines of the McDonnell-Douglas F-15. The T10-3 also incorporated aerodynamic changes based on the wind tunnel testing of T10-1 models.

The next move was a version called the T10-5, which also appeared in 1973. This differed from earlier configurations in having the engine accessory gearboxes located dorsally, which resulted in a marked decrease in maximum cross-section area. Then the T10-6A produced later in the year introduced some major changes. Like the T10-5, it was an integral-type layout with engines in spaced nacelles, a bubble canopy, twin tails and low-mounted stabilators. However, this project featured axis-symmetric air intakes that decreased the nacelle length and thus the overall surface area and drag. For the first time the main gear units retracted into fairings outboard of the engine nacelles so that the wheels lay horizontally in the wing roots. These fairings continued rearwards into beams, on which the tail unit was attached.

Also in 1973 the T-10 design team grew appreciably in size when a large number of engineers were reassigned to fighter aircraft from the terminated T-4 supersonic bomber programme (that aircraft had first flown in August 1972 but completed only ten flights). As a result the newcomers proposed an alternative blended wing/body configuration that was designated T10-6 and christened the 'power-pack' or 'package' layout. This was because the T10-6 had its engines located in a common nacelle with a single air intake divided into port and starboard halves by a vertical wedge; three-segment vertical intake ramps were attached to the wedge. The main

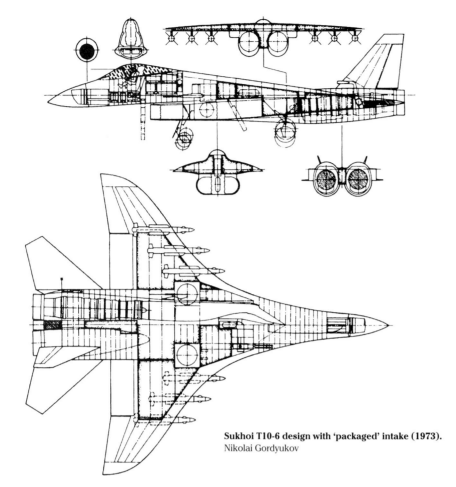

Sukhoi T10-6 design with 'packaged' intake (1973).
Nikolai Gordyukov

ture to the fighter's final configuration before the design was finally frozen in 1975. In fact the path taken to get to this point had brought considerable controversy and argument in regard to the aerodynamics of the ogive wings, and the T-10's overall integral layout. This had involved the Siberian Scientific Institute of Aerodynamics or SibNIA where T-10 wind tunnel models had revealed some unstable airflow that could produce buffeting in flight and create breakup of vortices. It then came to light that MAP's leaders had been misled by Sukhoi and had been given a rather favourable impression of the T-10 as originally designed, despite the fact that earlier SibNIA tunnel research had clearly shown airflow departure at an angle-of-attack of 10°. Tests with a model representing the full aircraft began in 1975 and these revealed poor control efficiency, roll and yaw stability.

During 1975 the OKB started issuing working drawings for the construction of the first prototype, which bore the manufacturer's designation T10-1 (not to be confused with the earlier T10-1 project designation). As the building work progressed it became clear that the T-10's layout and aerodynamics would be inadequate, but the aeroplane was duly completed because the Bureau wanted to test it in this form just in case a major redesign could be avoided. Lyul'ka's AL-31F engine was selected to power the new fighter. As noted earlier this was the engine designer's first afterburning turbofan and it was rated at 74.5kN (16,755lb) dry and 122.5kN (27,555lb) in reheat. The similar Solovyov D-30F6 afterburning turbofan, rated at 151.9kN (34,170lb) in full reheat, had also been considered but this was rejected because it was too heavy. However, since the new powerplant had not yet been cleared for flight test, the first T-10 prototypes were to receive AL-21F-3s to help accelerate their flight test programme.

In September 1975 Pavel Sukhoi died and shortly afterwards Chernyakov left for health reasons. The prototype T10-1 made its first flight on 20th May 1977 and a second AL-21F-3-powered prototype soon followed. Prototypes T10-3 (flown 23rd August 1979) and T10-4 were the same except that they introduced the AL-31F engine. It was immediately apparent that there were problems – in fact there were three main weaknesses. First, the avionics were heavier than originally predicted, which made the aircraft several hundred kilograms overweight; second, the AL-31F's fuel consumption was more

gear units retracted into fairings that flanked the nacelle and these continued aft into the beams on which were mounted the twin fins and low-mounted stabilators. The 'power pack' concept was clearly inspired by the T-4, which had four engines in a single nacelle and a similar wedge intake, but wind tunnel tests produced disappointing results and so the T10-6 was abandoned.

Each of these revisions added some fea-

Another late T-10 pre-project design shown in model form.

than expected (although the engine was
excellent); and finally, the ogival wings and
the position of the vertical tails on top of the
engine nacelles were not the best solution
and in a dogfight they would not give the
T-10 an advantage over US fighters. A full
redesign and a reduction in weight were
needed and, indeed, were demanded by the
VVS and MAP.

Designing the T-10's aerodynamics and
structure had taken nearly eight years and
there was no time left to try to invent a better
solution. The situation was so critical that
Yevgeniy A Ivanov (who succeeded Pavel
Sukhoi as General Designer) and Mikhail
Simonov (appointed T-10 project chief in
1976) urged their staff to bring in some new
ideas. Simonov took the brave and correct
decision to stop all work on the current aircraft, including plans for its production, and to
find an alternative. All of the previous hard
work was discarded and a sort of in-house
competition for the best layout began. The
task was to redesign the aircraft to meet the
VVS specifications but to stay within the current dimensions and weights. Virtually every
aspect of the design – the shape of the LERX,
the wing planform, the positions of the stabilators and the vertical tails – came under
review. SibNIA tested many versions of the
LERX in its wind tunnels in an attempt to find
a way of appreciably increasing lift while also
creating a sufficiently high pitchdown force at
any positive angle of attack. To give some
idea of the intensity of this work, before the
end of 1977 twenty-seven reports had been
published on efforts to refine the T-10's wings.

One positive benefit of having Simonov as
the T-10 project chief was that the pace of
work on the fighter's aerodynamics was substantially accelerated – his predecessors had
been particularly cautious. Simonov was
well aware that the basic features of the T-10,
the integral layout with an aft centre-of-grav-
ity, fly-by-wire controls and underslung
engines, would be retained but the aerodynamics needed a complete rework. Ivanov,
however, held a different view and still
hoped to get the original design to work as
originally thought, despite the fact that the
T-10's early test flights had revealed those
serious shortcomings which many believed
could not be overcome. In fact Ivanov was

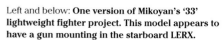

Left and below: **One version of Mikoyan's '33' lightweight fighter project. This model appears to have a gun mounting in the starboard LERX.**

Bottom: **The second Mikoyan '33' model.**

not the only opposition; the then Minister of Aircraft Industry Vassily Kazakov hit the roof at Simonov's proposal to redesign the T-10, denouncing it as an attempt to kill the project. This, of course, was common practice in the Soviet Union, where people would look for 'internal enemies' on whom they could then heap blame for failures and wrong decisions. Nevertheless, Simonov got his way and, after rejecting some alternative layouts (including forward-swept wings and canard foreplanes), in 1979 the 'T-10 Junior' entered full-scale development.

To distinguish it from the original T-10 the new design was designated T-10S, the suffix standing for *sereeynyy* (production); in fact T-10S bore only a slight resemblance to its predecessor. In order to reduce wing loading during take-off and aerial combat, the wing area was increased from 59.4m² (638.7ft²) to 62.04m² (667.1ft²), while the ogival wings gave way to a more traditional format with a straight leading edge outboard of the LERX and a new flatter aerofoil. To improve directional control at high alpha the fins were moved outboard to the stabilator attachment booms and mounted vertically, rather than being canted outwards as before. The combined effect of these changes reduced airframe drag by 18 to 20% in both the subsonic and supersonic regimes, while the overall maximum cross-section was also reduced and manoeuvrability much improved.

Design work was completed in 1980 and the first full T-10S prototype (coded T10-7 – the seventh T-10) made its maiden flight on 20th April 1981. The official service designa-

Soviet Secret Projects: Fighters Since 1945

tion Su-27 became known to the West in 1982 and the new type received the reporting name *Flanker*. In service the Su-27 has been a huge success and, as one might expect, this has led to many export orders and upgrades including multi-role types. They include the Su-27K (T-10K) naval carrier fighter that introduced a third horizontal flying surface in the form of canards fitted to the wing leading edge extensions – the full prototype first flew on 17th August 1987. Other versions to feature canards were the T-10M or Su-35 (an upgraded Su-27) and the Su-37 which had vectored-thrust engine nozzles. The latter first flew in this form on 2nd April 1996 and the presence of both canards and vectored thrust allowed it to perform some amazing aerobatics. Some production versions of *Flanker* now have vectored-thrust engines.

During 2003 plans were under way to modernise the Russian Air Force's Su-27's and to extend their service life until 2032. In service the Su-27P (the T-10S) has a service ceiling of 18,500m (60,696ft) and a rate-of-climb of up to 300m/sec (984ft/sec). Returning to the original T10-1 prototype, this aircraft carried no armament and, after completing its trials programme, it was left on the ground at Zhukovskiy. In 1985 it was donated to the Soviet Air Force Museum at Monino.

Mikoyan '33'

In about 1988 the Soviet Union reputedly offered a 'single-engined MiG-29' to India and in the early 1990s a TsAGI wind tunnel model and other artist's impressions were published in the West that showed a 'single-engine MiG-29'. During the late 1970s Mikoyan made some studies for a lightweight single-seat single-engine fighter and attack type but it is unknown if the two are the same. The late 1970s work came together under the designation 'Izdeliye 33', for which two different models have been found that exhibit only minor differences. Overall the '33' had a ventral intake and a single fin and looked to be something of a cross between the twin-engined MiG-29 and the American single-engined General Dynamics F-16 Fighting Falcon, although it was smaller than both.

The '33' had slim LERX with narrow chines from a vortex-generating claw alongside the forward cockpit and the intake was divided into two to allow the air to pass by the nose-wheel bay; the engine was single RD-33 as used on the MiG-29. One model has a more pronounced cockpit fairing and a sawtooth near the root of the tailplane, while the other appears to have a cannon mounted in the starboard wing leading edge extension; both have a pair of small ventral fins. A considerable amount of work was completed on the '33', including some tunnel testing on a 1.5m (4ft 11in) model, but no orders for a prototype were forthcoming. Part of the reason for this was that recent Air Force doctrine will not accept a single-engined fighter. The '33' label suggests that the aircraft may have been designated MiG-33, had it been built.

1970s Fighter Designs – Data / Estimated Data

Project	Span m (ft in)	Length m (ft in)	Gross Wing Area m² (ft²)	All-Up-Weight kg (lb)	Engine kN (lb)	Max Speed / Height km/h (mph) / m (ft)	Armament
Mikoyan MiG-31 (flown)	13.465 (44 2)	22.69 (74 5)	61.60 (662.4) without LERX	41,000 (90,388) Internal fuel only	2 x D-30F-6 90.8 (20,435), 152.0 (34,190) reheat	1,500 (932) at S/L, 3,000 (1,865) at 17,500 (57,415)	4 x R-33 + up to 4 x R-60 AAM or 3 x R-33 + 2 x R-40 AAM, 1 x 23mm cannon
Tupolev 'Aircraft 148' (Tu-148-100 System)	25.5 (83 8) minimum sweep 17.2 (56 5) intermediate sweep	31.2 (102 4.5)	?	55,000-60,000 (121,252-132,275)	2 x VD-19R2 139.1 (31,305)	1,400 (870) at low level, 2,500 (1,554) above 16,000 (52,493)	4 x K-100 AAM or various air-to-ground weapons (see text)
Tupolev 'Aircraft 148' (Tu-148-33 System)	26.6 (87 3) wing set at 20°	32.5 (106 7.5)	100.0 (1,075.3) wing set at 56°	60,000 (132,275)	2 x RD-36-41 156.8 (35,275) reheat	2,500 (1,554) at height with missiles	4 x K-33 AAM
Mikoyan 'heavy' MiG-29 (Single fin)	c10.1 (33 2)	c14.7 (48 3)	37.6 (404.3)	13,400 (29,541)	2 x R67-300 73.5 (16,534) reheat	1,500 (932) at S/L, 2,500 (1,554) at height	2 x K-25 and 4 x K-60 AAM, 1 x 30mm cannon (?)
Sukhoi T10-1 (1972 Conceptual Design)	12.7 (41 8)	18.5 (60 8)	48.0 (516.1)	18,000 (39,683) 21,000 (46,296) w tanks	2 x AL-31F 100.9 (22,710) reheat	1,400 (870) at S/L, 2,500 (1,554) at height	2 x K-25 and 6 x K-60 AAM, 1 x 30mm cannon
Sukhoi T10-2 (1972 Conceptual Design)	11.6 (38 1)	17.3 (56 9)	47.4 (509.7)	18,000 (39,683)	2 x AL-31F 100.9 (22,710) reheat	Over 2,000 (1,243) at height	2 x K-25 and 6 x K-60 AAM, 1 x 30mm cannon
Yakovlev Yak-45I	c9.8 (32 2)	c17.8 (58 5)	40.0 (430.1)	13,900 (30,644)	2 x R53F-300 80.4 (18,080) reheat	1,500 (932) at S/L, 2,500 (1,554) at height	2 x K-25 and 2 x K-60 AAM, 1 x 30mm cannon (?)
Yakovlev Yak-47	c12.2 (40 0)	c22.25 (73 0)	65.0 (698.9)	22,800 (50,265)	2 x R59F-300 122.5 (27,560) reheat	1,500 (932) at S/L, 2,500 (1,554) at height	2 x K-25 and 2 x K-60 AAM, 1 x 30mm cannon (?)
Mikoyan MiG-29 Early version (flown)	11.36 (37 3)	17.32 (56 10) with pitot	38.056 (409.2)	18,840 (41,534)	2 x RD-33 49.4 (11,110), 81.7 (18,385) reheat	1,480 (920) at S/L, 2,450 (1,523) at 11,000 (36,089)	4 x R-27R + 2 x R-60M or R-73, or 6 x R-73 AAMs. Various other stores. 1 x 30mm cannon
Sukhoi T10-1 Prototype (flown)	14.7 (48 3)	19.65 (64 5.5)	59.4 (638.7)	25,740 (56,746)	2 x AL-21F-3 76.4 (17,195), 109.7 (24,690) reheat	2,230 (1,386) at height	None carried but 8 hardpoints for 6 x R-27 and 2 or 4 x R-37 AAM. 1 x 30mm cannon
Sukhoi Su-27P (T-10S) (flown)	14.698 (48 3)	21.935 (71 11.5)	62.04 (666.7)	30,000 (66,138)	2 x AL-31F 74.5 (16,755), 122.5 (27,555) reheat	1,380 (858) at S/L, 2,500 (1,554) at 11,000 (36,089)	R-27 and R-73 AAM or various other bombs and stores on 10 hardpoints. 1 x 30mm cannon

Chapter Ten

Recent and Future Programmes

Since the break-up of the Soviet Union after the end of the Cold War in 1990/91, the acquisition of new combat aircraft (and indeed all forms of military equipment) has been an extremely difficult proposition for the Russian nation. Prior to 1990 the Soviet armed forces had considerable control over the supply of money that they received plus a large volume of labour to back up the acquisition of their aircraft. Since 1990 they have been starved of funds with the result that the flow of new fighters has almost ground to a halt and some promising new designs have had to be abandoned. During the Cold War the Soviet Air Forces were acquiring around four hundred new aeroplanes a year, yet by 1992 the VVS could afford just thirty-two, and none at all were bought in 1997. In addition, flying hours and training were reduced to a bare minimum, enormous numbers of staff were sacked and many old and not so old aircraft

were scrapped. To cut costs and reduce duplication, in 1998 the PVO and VVS were merged and hundreds of operational units from both forces were disbanded. Only recently has progress been made on producing some new fighters.

Incredibly, by 1998 all of the OKBs, production plants and research institutes that had been involved with fighter production were still active, but with an output of less than 10% of their former level. There is currently a substantial need for new and modern combat aircraft and equipment and plans are now in place to get them, but only if the finance is there. During 2002 and 2003 the number of flying hours began to rise as some extra money began to trickle through and it is hoped that funding will continue to build up in the near future. The fighter projects that have been under consideration during the fifteen years since the end of the Cold War are as follows.

Although it failed to gain any production orders, the Mikoyan 1-44 prototype was a very impressive aeroplane.

MFI Programme

In the late 1970s research work began on a new fifth-generation fighter under a programme called the I-90 (*istrebitel* or fighter for the 1990s) which, combined with efforts to find new bomber and attack types as well, would require a massive research and development effort. As one might expect the various aviation research institutes were heavily involved with both the fighter and with the new weapons and avionics that would be required to go in it. The first complete designs began to appear in the early 1980s and, in common with new fighter proposals around the rest of the world, the I-90 had to be very

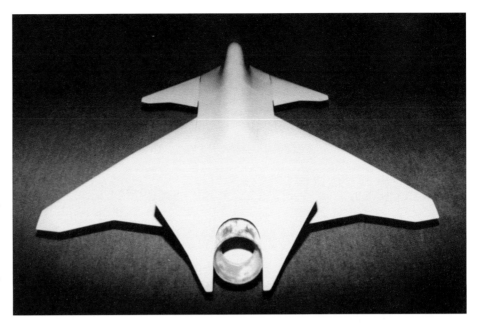

manoeuvrable and was to incorporate stealth technology. It also had to be capable of super-cruise: the ability to fly supersonically without using engine reheat. This was a major step forward because, apart from the MiG-31 whose Solov'yov D-30-F-6 units were opti-mised for prolonged supersonic cruise, none of the USSR's fourth-generation fighters could go supersonic without using their afterburn-ers. Consequently this increased their rate of fuel consumption to a very high level and to a point where they could remain supersonic for no more than ten to fifteen minutes. The new requirement called for much longer periods of time above Mach 1 but with the use of reheat restricted to situations where a rapid acceleration or sharp manoeuvres were needed.

Following experience with the Su-27 and MiG-29, it was hoped to develop two parallel types, a heavy twin-engined long-range fighter with powerful weapons and radar and a lightweight single-engined tactical fighter, the two aeroplanes employing the maximum possible commonality. The heavy type was duly designated MFI for *mnogofoonktsion-ahl'nyy frontovoy istrebitel* or multi-function fighter, its lightweight cousin was the LFI for *lyohkiy frontovoy istrebitel*. All three fighter OKBs eventually offered designs although, to begin with, Yevgeniy Ivanov (the General Designer at Sukhoi) had declined to take part because his Design Bureau felt that the latest generations of Su-27 would more than suf-fice. However, once Ivan Silayev, the Minister of Aviation Industry, had insisted that Sukhoi should participate, the Bureau responded with a forward-swept wing design. Concur-rently the Lyulka-Saturn Design Bureau was to produce a new engine specifically tailored to fit the new fighter and this eventually appeared as the AL-41. The full-scale devel-opment of this power unit was approved by a directive issued in 1986.

Yakovlev MFI

This single-engined canard delta design fea-tured outwardly canted twin fins and, unusu-ally, a kinked trailing edge on both the canard and the main wing. It had side intakes placed beneath the canards and the forward fuselage and chines were not unlike those on the new American Lockheed Martin F-22 Raptor stealth fighter. The engine nozzle was capable of a degree of vectored thrust because it could move up and down in the vertical plane. Dur-ing the decision-making process the design's

single engine powerplant counted against it because, for over twenty years, the Soviet Air Force had favoured having at least two engines in multi-role aircraft. Thus the single power unit was considered to be a big weak-ness and so, once again, a Yakovlev fighter design received an official rejection.

Sukhoi MFI/S-32

As noted, Sukhoi's MFI offering had a for-ward-swept wing, which in fact benefited from some new aerodynamic research recently completed at TsAGI and SibNIA. It was intended that the aircraft should possess outstanding agility and it was to be capable of controlled flight at very high angles of attack up to 90° or even more. After Mikoyan had been declared the winner Sukhoi's project was continued as a private venture because the basic design had showed some promise –

the subsequent work is described shortly. Ini-tially the aircraft was given the in-house title S-32, although it was later to receive two fur-ther designations.

Mikoyan 1-42 and 1-44

In the early 1980s Mikoyan worked on designs to fulfil both the MFI and LFI parts of the requirement, the respective designs being known as the 5-12 and 4-12 (and also appar-ently 1-41 and 1-43). They were intended to share as many components as possible. After further research TsAGI's experts recom-mended that Mikoyan should choose a design fitted with a canard and a delta wing and in 1987 the completed advanced devel-opment projects for the two requirements were submitted for review within the official competition. Both were accepted and thus progressed into the full-scale development

The near-complete Mikoyan 1-44 prototype seen in the Design Bureau's experimental shop.

important factor because Mikoyan's project was thought to offer a smaller risk than Sukhoi's forward-swept wing.

The MFI label meant that the aircraft would need to carry both air-to-air and air-to-ground weaponry and the internal carriage of this ordnance (to keep down the fighter's radar cross section) absorbed a great deal of thought and effort. Mikoyan's design also had a mass of control surfaces to assist the aerodynamics but certain features, like the twin fins and dorsal spine, did not help the design's stealth characteristics. The 1-42 possessed a fly-by-wire control system and the avionics were the most advanced yet produced in the Soviet Union. The first prototype was officially classed as a technology demonstrator and was thus given a separate designation 1-44; as such it was intended to prove the design's aerodynamics and control systems. Just how close the 1-44 is or was to the ultimate 1-42 layout is still unknown because few details of the latter have been released. One assumes that experience from the flight test programme would have been worked into the 1-42 and would most probably have brought some internal and external changes.

By 1994 the 1-44 airframe was complete and nearly ready for its first flight, but it still lacked some of its equipment. In particular the flight control actuators were absent because ANPK Mikoyan (a new name which had recently been given to the OKB) had been unable to pay the manufacturers of this equipment. The squeeze on funding that became so widespread during the 1990s had begun to bite and in fact, for the 1-44, the money was to dry up completely. In December Rostislav Belyakov announced that the

phase, but the Bureau soon found that it could not finance the two programmes simultaneously. The LFI was shelved almost immediately and from here on in the design team (led by Gheorgiy Sedov) concentrated on the heavy MFI. By now this had been redesignated 'Izdeliye 1-42'.

The choice of Mikoyan's project as the winner was, in part, a political decision because

Mikoyan was now considered to be, quite simply, a highly specialised 'fighter designer'. In comparison Sukhoi had expanded into a much broader area having worked towards military requirements for both bombers and attack types, while Yakovlev was also involved with civil aircraft; as such it was not quite so critical for Sukhoi to win the competition. The technical aspects, however, had also been an

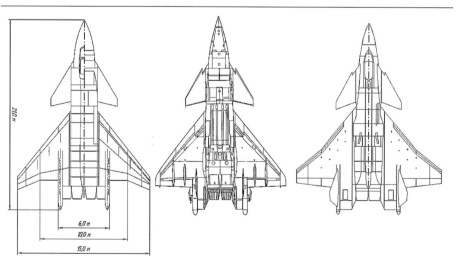

6,0 м
10,0 м
15,0 м

It seems a pity that the 1-44 completed only two test flights. This picture was taken during the first flight when the undercarriage was not retracted.

This set of drawings is believed to show the evolution of the Mikoyan 1-44, although their source is uncertain.

aircraft was being prepared for a maiden flight that would take place 'in the near future' and fast taxi trials were undertaken during that month. Nevertheless, despite several reports claiming that flight was imminent, the aircraft stayed on the ground (and out of sight to all observers) for a long time. Finally, in early 1999 the first photograph was released to the public and on 12th January an official roll-out was made for the press, the event including a taxi along the Zhukovskiy runway. On 29th February 2000 the 1-44 at long last made that maiden flight and a second sortie followed on 27th April, but apparently the aircraft has not flown since.

The reasons for this are clear. During the time the 1-44 spent on the ground it has been left behind by technical progress, while the 1-42 programme has been replaced by a new fifth-generation fighter described later. In 1999 there were still plans to buy the 1-42, but it was quickly realised that Russia could not afford to acquire enough examples to form an operationally useful force, so more airframes from the Su-27 family were bought instead. Full data for the 1-44 has never been released and during the two flights that were made the aircraft achieved only modest speeds and a height of just 2,000m (6,562ft). Had it entered service the 1-42 would most likely have been given the designation MiG-37 or MiG-39; curiously, the aircraft was recently given the western codename *Flatpack*.

Sukhoi S-32/S-37/S-47 Berkoot

It was the OKB's General Designer Mikhail Simonov who did much to keep Sukhoi's 'MFI' project going although, fortunately for the Bureau, overseas sales of its Su-27 family also kept sufficient cash coming in to allow the development and manufacture of this new prototype to proceed. The staff realised the value of keeping up to date with the advanced technology required to design new fighters and, had they let this opportunity go, the Bureau might never have caught up in the new and very competitive post-Soviet Union era. The forward-swept wing (FSW) offered many advantages including a better lift/drag ratio in manoeuvres, longer subsonic range, better control at low subsonic speeds and higher lift. The aeroplane was also near spin-proof. Besides being called S-32 (a number used previously by a version of the Su-17 *Fitter* fighter-bomber) the machine was also given a name – Berkoot (Golden Eagle), a most unusual step in Russian fighter aircraft

history. In fact the S-32 was really an FSW ultra-manoeuvrability technology demonstrator and numerous changes were made to its design and its aerodynamics as the number of wind tunnel hours grew. Chief project engineer was Mikhail Pogosyan.

It is believed that, for trial purposes, Sukhoi had previously fitted a forward-sweep wing to an Su-9 airframe. In addition an FSW demonstrator aircraft (coded SYB-A in the West) was flown in about 1982 and this may have been the same aircraft. The S-32 itself was intended to receive two Khachatoorov R79M-300 turbofan engines each giving 181.3kN (40,785 lb) of thrust in reheat. This was a version of the Yak-41 V/STOL aircraft's R79V-300 lift/cruise engine fitted with a different nozzle that could be deflected +/- 20° in the vertical plane, but

the powerplant was eventually replaced by the Lyul'ka-Saturn AL-41 (in fact two Solov'yov D-30F-6M units were to be fitted for the aircraft's first flights). In the early stages of the programme the S-32's layout had featured a single two-dimensional nozzle that served both engines – a unique feature in aircraft design. However, tests showed that in its current form the S-32 would be overweight and incapable of achieving the performance requirements set for it by the Air Force. As a result a substantial redesign followed and the aircraft was redesignated S-37.

When the prototype's construction began in the late 1980s it was envisaged that two examples plus a static test airframe would be produced. However, the shrinking defence budget cut this to just the one airframe and,

The Sukhoi S-37 Berkoot forward-swept-wing technology demonstrator.

as noted, Sukhoi eventually had to find its own money to complete and get the S-37 into the air. The S-37 made its maiden flight on 25th September 1997 and, unlike the Mikoyan 1-44, it has completed a substantial amount of flying which has continued well into the new century. Unofficial reports have stated that it has achieved a speed of Mach 1.3, the highest ever recorded by a forward swept wing aeroplane. No weapons have been carried although a combat version would receive a single GSh-301 30mm cannon mounted in the wing roots plus a selection of air-to-air missiles. The S-37's service ceiling is 18,000m (59,055ft).

In 2000 the aircraft was redesignated S-47 to prevent confusion with the Su-37 vectored thrust version of Sukhoi's *Flanker* (Chapter Nine). It has also received a Western codename – *Firkin*. Like Mikoyan's 1-44, this aircraft will not be developed further and will never enter Russian Air Force service. Instead a new fighter from Sukhoi called the T-50 is being built and this will be discussed later, but the input of Berkoot experience into the new type will be substantial.

Other Projects

Sukhoi S-37

Prior to its application to Sukhoi's FSW demonstrator, the S-37 designation was previously carried by this fighter-bomber design which was first unveiled to the public at a trade air show held at Dubai in early November 1991. A canard delta design of a form then popular with Western fighter designers of the Eurofighter and French Rafale, this was intended to be a true multi-role combat aircraft carrying either one or two crew (models were displayed in both forms). Previously types like Sukhoi's Su-25 *Frogfoot* attack aircraft had been tailored to specific missions. The S-37's basic roles were to strike surface targets in any weather conditions day or night and air superiority operations well beyond the front line; it was also expected to serve as an interceptor and as a reconnaissance aircraft.

A high level of agility was called for together with a good low-level transonic attack performance and minimal radar cross section, factors which meant that a layout with large podded engines had to be rejected. Therefore the S-37 featured close-coupled foreplanes with the intakes underneath and wing

leading edge extensions above. The canards were stated to have a deflection range of +10° to -70° to give better low-speed control in pitch. The intakes had no moveable parts, which meant that speeds of Mach 2 could be achieved, but the intakes themselves would still retain their effectiveness at very high angles of attack. The wings had leading edge slats and trailing edge elevons and could be folded to allow stowage in small areas, with a folded span of 8.10m (26ft 7in). The powerplant was to be a single AL-41F engine.

A fly-by-wire control system was to be fitted together with a pulse Doppler low-altitude terrain-following radar that was suited for transonic speeds but which could also perform air-to-air intercept duties. In fact the former Soviet Union's deficiencies in avionics and electronics would hopefully have been addressed by the equipment in this aircraft. The radar was expected to deal with land and sea surface targets and to locate, track and destroy low-level flying targets at all speeds. These included a hovering helicopter, which would be difficult to identify against ground clutter. Altogether, the radar was expected to be able to track ten targets simultaneously.

Based on experience in the Afghan War

Sukhoi S-37 multi-role fighter project (1991).

Views of a model of the Sukhoi S-37. George Cox

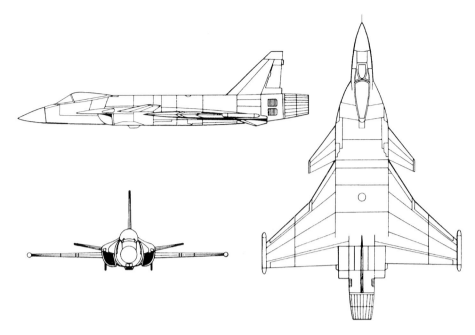

with the Su-25, up to 800kg (1,764 lb) of armour had been worked into the airframe to protect the pilot, powerplant and some critical airframe sections. A single GSh-30 30mm gun was mounted in the starboard wing root extension and there were no less than seventeen (or eighteen) hardpoints for weapons – nine (or ten) beneath the fuselage and eight more under the wings. These were intended to take all manner of stores, embracing laser and TV-guided air-to-surface missiles, anti-radar missiles, short- and medium-range air-to-air missiles, anti-tank missiles, unguided rocket pods, guided, retarded or conventional bombs (up to a maximum size of 1,500kg [3,307 lb]) and, finally, podded 30mm guns. Various reconnaissance pods could also be carried and the maximum external war load was 8,000kg (17,637 lb) and maximum fuel load 8,300kg (18,298 lb).

The S-37 was designed to operate from unpaved surfaces, it was stressed for -3G and +9G and was to be capable of 8G manoeuvres during supersonic flight. The canard delta wing plan was seen as the aerodynamic ideal to match the various mission requirements and with its fuselage design the S-37 was optimised for operations at low altitude. An exceptional range/payload was possible which could be increased rather more with in-flight refuelling – the estimated combat radius with a 3,000kg (6,614 lb) load was 1,500km (932 miles). Service ceiling was 17,000m (55,774ft) and, when flying at an altitude of 1,000m (3,280ft), the S-37 was expected to accelerate from 600km/h (373mph) to 1,100km/h (684mph) in 14 seconds and from 1,100km/h to 1,300km/h (808mph) in 7.2 sec-

The Mikoyan 701 is believed to have looked very similar to this model. John Hall

onds. Such a performance would require a very powerful engine.

Potentially this was a very capable aircraft. The S-37's chief designer was Vladamir Bobak who had also worked on the subsonic Su-25 attack aircraft described in the companion volume on bombers; in fact the S-37 was sometimes presented as a successor to the Su-25. Bobak believed that the demand for an S-37-type of aeroplane was growing but by 1992 the project was stalled due to a lack of funding, even though it was still only at the basic design stage with wind tunnel testing under way (some of the systems, including the radar, engine and software, had also apparently been tested). Attempts were made to find a foreign partner who could offer further investment but these proved fruitless and so work on a prototype was never started.

Despite some comments to the contrary, neither the Russian government nor the Air Force showed much interest in the project and it was effectively terminated during 1992. The fact that it was a single-engined machine may not have helped because, as noted previously, the Russian Air Force had a preference for multi-engined combat aircraft. Such types were less vulnerable to battle damage and, since the S-37 was expected to operate frequently at low level, its vulnerability would therefore have been that much higher. In fact Sukhoi did consider making the S-37 a twin-engined aircraft but it seems that the single-engine layout would have been cheaper to produce.

Mikoyan 701

This is still very much of a mystery aeroplane because, to date, few details and no accurate illustrations have been released. It was first proposed in the late 1980s to replace the MiG-31 in the ultra-long-range interceptor role. To help protect Russia's Northern and Eastern regions the aircraft was expected to have a range of 7,000km (4,350 miles) when cruising at speeds between 2,300km/h (1,430mph) and 2,500km/h (1,554mph) and at heights of around 17,000m (55,774ft), while the range when flying subsonically would rise to about 11,000km (6,835 miles). The sketches that have been published suggest that the 701 was a long sleek aircraft with small canards, a large 'cranked' delta wing with substantial leading edge extensions and twin jet engines mounted in the upper rear fuselage. These were fed by a large split dorsal intake and a small fin sat on top of the engine nacelles. Estimated span was 19m (62ft 4in) and length 30m (98ft 5in) and there are no clear details of weapons, but one expects the armament to have been substantial and it would most probably have been carried internally. It seems likely that the 701 fell victim to the cash shortages of the 1990s although some years ago an announcement stated that it was still an active project.

Lightweight Studies

Sukhoi S-54

This design began life as Sukhoi's contender to a competition launched in 1990 to find a new advanced two-seat jet trainer and light combat aircraft. It had a swept wing and tail, twin outward-canted fins, intakes underneath the wing roots, and was powered by a Soyuz/Tumanskiy R-195FS turbojet which gave 41.2kN (9,270lb) of thrust dry and 60.8kN (13,680lb) with reheat. Six air-to-air missiles could be carried on wingtip and four underwing hardpoints and the estimated top speed was 1,200km/h (746mph) Mach 0.98 at sea level and 1,650km/h (1,025mph) Mach 1.55 at height. The design was refined in 1992 and had a span of 9.08m (29ft 9½in), length 12.30m (40ft 4in) and a maximum take-off weight of 9,410kg (20,745lb), but it failed to win the trainer competition.

However, in 1996 a new version using an airframe that was about 25% larger was unveiled at that year's Farnborough Air Show. Yet another quite different configuration was displayed at the June 1997 Paris Air Show, by which time the project had become a fully-fledged light fighter with canards, vertical mounted fins and a chin intake. It was now equipped with a Phazotron Sokol-X radar, which suggested a change of emphasis to the combat role with training now reduced to a secondary capability. Little data has been revealed for this later version and in more recent times the promotion of this aircraft appears to have come to an end. Sukhoi described the S-54 as a scaled-down single-seat single-engined development of the Su-27 with speed, manoeuvrability and operating altitude on a par with current combat aircraft requirements. The fighter would have employed a fly-by-wire control system and its wingspan is believed to have been 11.23m (36ft 10in) and its length 15.30m (50ft 2½in).

Model of the Sukhoi S-54 in its ultimate configuration.

Sukhoi S-55 and S-56

The lightweight S-55 fighter project emerged in the mid-1990s and was a development of the S-54 produced specifically for export. It was a tandem-seat two-seater and had canards, a swept wing, swept tail and out-ward-canted twin fins. The weapon system was based on the vectored thrust Su-37 (Chapter Nine) and included an N011M radar plus the usual range of air-to-air missiles (models of the S-54 and S-55 were both seen with R-73 AAMs on the wingtip stations and R-77 AAMs and Kh-29 air-to-surface missiles on the wing pylons). Power was to come from a single thrust-vectoring Saturn AL-31 or AL-37 engine: the estimated service ceiling was 18,000m (59,055ft) and range on internal fuel 820km (510 miles) at sea level or 3,000km (1,864 miles) at height. During the summer of 1995 a decision was taken to build a prototype of the S-54/S-55 with a first flight due in the late 1990s, and in mid-1996 the mock-up was near-complete, but nine years later it appears work is no longer ongoing. Sukhoi's S-56 project was also proposed in the 1990s and is believed to have been a car-rier-based lightweight fighter, again with exports in mind. It is also believed that the design was offered to India in November 1999.

PAK FA Programme

LFS Projects

Alongside Sukhoi's private projects, the LFI programme noted earlier was revived in 1994 as the LFS and Mikoyan, Sukhoi and Yakovlev all became involved. However, LFS stood for *Lyogki Frontovoy Samolyot* or Lightweight Tac-tical Aircraft (rather than Fighter), which indi-cated that it would be another multi-role type. Designs were sought from all three Bureaux with the intention of choosing one to fill the future fifth-generation fighter requirement. Sukhoi offered the S-54 and also, it is believed, a scaled-down S-37 (S-47) powered by an AL-41 engine and with the forward-swept wing substituted by a more conventional alternative. The latter was under way by 1998 and was apparently to be powered by two RD-133 engines or a single AL-31PF, both versions also being equipped with thrust vectoring.

Mikoyan apparently considered both single and twin-engined projects and revealed the existence of its studies in the spring of 1997. The preferred twin-engined version would use

98kN (22,050 lb) thrust engines to give a power/weight ratio of 1.3:1; the single-engined variant's equivalent ratio was to be 1.1:1. Around 2000 Yakovlev proposed a STOVL project with some stealth features that was clearly based on the Yak-41/Yak-43 family (Chapter Six). However, from the outset the then head of the Yakovlev Bureau, Mr Dondukov, was sceptical about the chances of obtaining Ministry funding for his team's project.

Sukhoi T-50

The original Russian Defence Ministry specification for a new fifth-generation fighter was drafted in 1998 but the plans were refined in 2000/2001 and, as a result, a new competition opened between the two main players – this was now the official replacement for the MFI programme that had produced the Mikoyan 1-42. A revised specification called for a 'super agile' 20,000kg (44,092 lb) class fighter with a low radar signature and the capability to cruise at supersonic speeds. Up to eight air-to-air missiles would be carried and a brand new radar was to be designed for the aircraft by the Tikhomirov research institute. The biggest change was that the project was no longer a lightweight type but rather a medium-size fighter which would fall roughly between the MiG-29 and Sukhoi Su-27.

During 2001 both OKB Sukhoi and RSK MiG (Mikoyan's name from 1999) submitted detailed proposals to the Defence Ministry Tender Committee. They both described a combined consortium that included other companies and research establishments to control all of the various elements of the project. Mikoyan's proposal was a quite different machine to the earlier 1-42, being tailless and much lighter and presenting a considerable

technical risk; its take-off weight was stated to be 19,000kg (41,887 lb). Sukhoi's project was a further development of the S-47. Yakovlev did not submit a design of its own but instead was specified as a sub-contractor to both competitors to provide the STOL and vertical landing elements for their fighters, areas in which Yakovlev specialised (in fact within Russia Yakovlev has no competition in this field). Yakovlev military aircraft chief designer Konstantin Popovitch announced at the ILA'02 Show at Schoenefeld that his company estimated its share in the programme to be as much as 30%.

In early May 2002 it was announced that Sukhoi was the winner but the Ministry-Industry Committee requested that both RSK MiG and Yakovlev should participate in the overall programme. In fact the winning partner had to guarantee that the losing contractors would contribute and so Sukhoi has design leadership but Mikoyan and Yakovlev retain their involvement. At the time Sukhoi's general director, Mikhail Pogosyan, also declared that foreign partners would probably be needed to help share the cost. The primary contenders here are China and India although the Ukraine is also apparently being considered. By now the Air Force had retitled the project PAK FA, the *Perspektivnyi Aviatsionnyi Kompleks Frontovoi Aviatsyi* or Prospective Aviation Complex for Frontal Aviation, and the new fighter was intended to replace both the MiG-29 and Su-27.

Following the financial problems suffered by Russia during the 1990s, finding sufficient funding to build enough T-50 airframes will be a substantial exercise. To date no official drawing has been released but provisional sketches suggest an airframe that has much in

common with the S-47 Berkoot, but with the forward-swept wing replaced by a more conventional trapezoid form; there is also a swept tailplane. Wingspan is roughly 15.5m (50ft 10in), length 23.0m (75ft 5½in) and maximum take-off weight 33,000kg (72,751 lb). Two vectored-thrust NPO Saturn AL-41F1 engines are to be fitted giving 151.9kN (34,170 lb) of thrust each. (Recently the three main Russian engine manufacturers, Rybinsk, Lyul'ka/Saturn and UMPO, have been merged into a single entity called NPO Saturn). To assist the fighter's stealth characteristics, all of the weapons are to be carried within the fuselage. In early 2004 the planned timetable for the T-50 was a maiden flight in 2006 with the type due to enter production in 2010. In December 2004 the technical specification was revised again with the maximum speed being reduced from 2,500km/h (1,554mph) to Mach 2; supercruise speed is set at around Mach 1.6.

This is the first large-scale fighter programme to emerge from post-Soviet Union Russia since the end of the Cold War and it means a great deal to the future of the country's aircraft industry. The VVS and the various design bureaux were unanimous that such a project should get under way to keep the development of advanced military aircraft up to date and their technical expertise on a par with the rest of the world. At the time of writing (mid-2005) it would seem that the Sukhoi organisation is now the strongest of the Soviet fighter manufacturers. Yakovlev currently has other trainer and civil work but Mikoyan does not appear to have too much in the way of fighter orders and employment outside the PAK FA, so in the long term one wonders if this famous name will survive. Only time will tell.

1990s Fighter Designs – Data / Estimated Data

Project	Span m (ft in)	Length m (ft in)	Gross Wing Area m² (ft²)	All-Up-Weight kg (lb)	Engine kN (lb)	Max Speed / Height km/h (mph) / m (ft)	Armament
Mikoyan I-44 (flown)	c16.3 (53 6)	c21.7 (71 2.5)	c90.5 (973.1)	c35,000 (77,160)	2 x AL-41F 173.4 (39,020) reheat	2,500 (1,554) at height	R-77 and R-73 AAM, 1 x 30mm cannon
Sukhoi S-37/S-47 Berkoot (flown)	16.7 (54 9.5)	22.6 (74 2) including pitot	56.00 (602.15)	34,000 (74,956)	2 x D-30F-6 196.0 (44,090) reheat	Mach 1.3 achieved at height. Estimated max: 1,400 (870) at S/L, Mach 1.6 at height	None carried
Sukhoi S-37 (Single-engine)	11.80 (38 8.5)	17.50 (57 5)	50.0 (537.6)	25,000 (55,115)	1 x AL-41F 181.2 (40,775) reheat	1,500 (932) Mach 1.22 at S/L, Mach 2 at height	Up to 8,000kg (17,637 lb) of stores, 1 x 30mm cannon
Sukhoi S-55	9.59 (31 5.5)	13.24 (43 5.5)	?	?	1 x AL-31 or AL-37	1,200 (746) at S/L, 1,650 (1,025) at height	Various AAM

Soviet Secret Fighter Colour Chronology

Top: **Mikoyan SM-12 research prototype.**

Centre: **Mikoyan Ye-5.**

Bottom: **Mikoyan I-420/I-3U.**

Opposite page:

Top: **Mikoyan I-7U.**

Centre left: **Mikoyan Ye-8 first prototype.**

Centre right: **Sukhoi Su-15.**

Bottom: **Twin-engined Mikoyan Ye-152A prototype.**

This page:

Top: **Model of the Mikoyan Ye-152P/M project with canard foreplanes.**

Centre: **Mikoyan 23-31 VTOL research prototype.**

Bottom: **Mikoyan 23-01 prototype STOL fighter.**

This page:

Top: **Mikoyan 23-11 swing-wing prototype which led to the MiG-23.**

Centre: **Three different Mikoyan Ye-155 project models showing designs which formed part of the research leading to the MiG-25.**

Bottom left: **Model of the swing-wing Ye-155MP project which helped to bring forth the MiG-31.**

Bottom right: **Fixed wing Ye-155MP project.**

Opposite page:

Top left: **Mikoyan's 518-55 project was another to be studied during the lead up to the MiG-31.**

Top right: **Model of a Yakovlev Yak-41 pre-project.**

Centre: **Mikoyan MiG-31.**

Bottom: **Yakovlev Yak-41 prototype.**

Above and right: **Tupolev Tu-148-100 long-range interceptor with its variable-geometry wings in their minimum and maximum sweep positions.** John Hall

Model of the Sukhoi T-37 heavy fighter. John Hall

Model of the Sukhoi Su-19M. John Hall

Model of the 'heavy' Mikoyan MiG-29 project.

Sukhoi's 'conventional' T-10-2 design with box intakes. John Hall

Soviet Secret Projects: Fighters Since 1945

The first Sukhoi T-10 (Su-27) prototype.

First prototype of the Mikoyan MiG-29.

The sole Sukhoi Su-37 prototype was essentially a standard Su-27 fighter fitted with vectored-thrust engine nozzles, which in this view can be seen facing downwards.

Mikoyan 1-44 technology demonstrator prototype.

Glossary

AAM Air-to-air missiles.

AI Air Interception.

Anhedral Downward slope of wing from root to tip.

AoA Angle of attack, the angle at which the wing is inclined relative to the airflow.

Angle of Incidence Angle between the chord line of the wing and the fore and aft datum line of the fuselage.

Area Rule Principal law for keeping transonic drag to a minimum. States that cross-section areas of aeroplane plotted from nose to tail on a graph should form a smooth curve.

ASCC Air Standards Co-ordinating Committee.

Aspect Ratio Ratio of wingspan to mean chord, calculated by dividing the square of the span by the wing area.

Chord Distance between centres of curvature of wing leading and trailing edges when measured parallel to the longitudinal axis.

CofG Centre of gravity.

Council of Ministers A Committee formed from those Ministries directly involved with the Soviet Aviation Industry – Ministry of Aircraft Industry, Defence, Civil Aviation, Higher and Specialised Education and Foreign Trade.

Critical Mach Number Mach number at which an aircraft's controllability is first affected by compressibility, that is, the point at which shock waves first appear.

Dihedral Upward slope of wing from root to tip.

Fly-by-Wire Flight control system using electronic links between the pilot's controls and the control surface actuators.

ECM Electronic Countermeasures.

GAZ *Gosudarstvenny Aviatsionny Zavod* (State Aircraft Factory).

GKAT *Gosudarstvenny Komitet Aviatsionny Teknniki* (State Committee for Aviation Equipment).

GKO *Gosudarstvenny Komitet Oborony* (State Committee for Defence).

GosNIIAS State Research Institute of Aviation Systems.

ICBM Intercontinental Ballistic Missile.

IFR In-flight refuelling.

Istrebitel' Fighter.

Izdeliye Product, Article or Aircraft.

Laminar Flow Wing Specifically designed to ensure a smooth flow of air over its surfaces with uniform separation between the layers of air.

LERX Leading edge root extensions.

LII Flight Test Research Institute of the Ministry of Aircraft Industry (MAP).

Mach Number Ratio of aeroplane's speed to that of sound in the surrounding medium – expressed as a decimal.

MAI *Moskovskii Aviatsionii Institut* (Moscow Aviation Institute).

MAP *Ministerstvo Aviatsionnoy Promysh Lennosti* (Ministry of Aircraft Industry).

NATO North Atlantic Treaty Organisation.

NII VVS *Nauchno Issledovatelyskii Institut* (Soviet Air Force Scientific and Research Institute based at Akhtoobinsk).

NKAP *Narodny Komissariat Aviatsionnoi Promyshlennosti* (People's Commissariat for Heavy Industry).

NTS Scientific and Technical Committee.

OKB *Opytno Konstruktorskoye Byuro* (Experimental Construction/Design Bureau).

Politburo The Central Committee of the Communist Party. This was responsible for policy and worked with the Council of Ministers.

PVO *Protivovozdushnaya Oborona* (Air Defence Forces).

RATOG Rocket-Assisted Take-Off Gear.

SibNIA Siberian Scientific Institute of Aerodynamics

S/L Sea level

SKB Special Design Bureau.

SovMin Soviet Ministry.

STOL Short Take-Off and Landing.

STOVL Short Take-Off and Vertical Landing.

t/c Thickness/chord ratio.

Transonic Flight The speed range either side of Mach 1.0 where an aircraft has both subsonic and supersonic airflow passing over it at the same time.

TsAGI *Tsentrahl'nyy Aero-i Ghidrodinameecheskiy Inst‌itoot* (Central State Aerodynamic and Hydrodynamic Institute), Zhukovsky.

TsIAM Central Institute of Aviation Motors.

VG Variable Geometry

VIAM *Vsesoyuzny Institut Aviatsionnykh Materialov* (All-Union Institute for Aviation Materials).

VRD *Vozdooshno-Reaktivnyy Dvigatel* (Air Reaction Engine).

V/STOL Vertical/Short Take-Off and Landing.

VTOL Vertical Take-Off and Landing.

VVS *Voenno-Vozdushniye Sily* (Air Force of the USSR).

Soviet and Russian
Fighter Project Summary

This list embraces all known Soviet and Russian post-war fighter projects plus any other research aircraft specifically intended to help and advance the art of the fighter designer. Care must be taken with some designations because the re-use of project numbers was a common and infuriating habit of Soviet and Russian OKBs. In particular the order of the projects emanating from the Sukhoi OKB is very difficult to decipher.

In the period around the end of the Second World War the Lavochkin, Mikoyan, Sukhoi and Yakovlev OKBs (Experimental Construction Bureaux or Design Bureaux) specialised in the design of fighter aircraft while Ilyushin and Tupolev worked on multi-engined bombers. Ground-attack aeroplanes were also the domain of Ilyushin and, in addition, for a short period during the late 1940s the Alekseyev OKB also produced a series of fighter designs and prototypes. In 1960 Lavochkin became involved with spacecraft and missiles, which left the other three fighter specialists to satisfy the needs of the Soviet Armed Forces and this situation changed relatively little until after the end of the Cold War in 1990. Nevertheless one of the bomber specialists, Tupolev, did produce a small number of fighter projects and one of these, the Tu-28, served in the front line for a long time.

Mikoyan devoted almost all of its efforts towards fighters although a few designs were adapted for ground-attack work. During the 1960s however, Sukhoi, after its Su-7 became a successful fighter-bomber, diversified into bomber development with the Su-24 strike aircraft and then the T-4 strategic bomber prototype. In contrast Yakovlev's interests stretched across a variety of areas, both in the civil and military sector. Following its successful Yak-25/28 series, Yakovlev's later efforts in fighter design were concentrated on vertical take-off, although a small selection of conventional designs still emerged from its offices. Some of the bureaux enjoyed an uninterrupted existence throughout the period covered by this book, although the financial restraints and lack of orders in the 1990s increased the necessity for them to work together so that they could compete in the international market place. Since the mid-1990s both Mikoyan and Sukhoi have operated under modified names to reflect their changing economic and business situations.

From 1954 a body called the Air Standards Co-ordinating Committee (ASCC), made up of representatives from America, the United Kingdom, Canada, Australia and New Zealand, introduced a series of 'reporting names' for Soviet aircraft. The idea was to assign a supposedly unambiguous name that could be easily identified over a poor voice radio link. Fighters were represented by F-names, bombers by B, air-to-air missiles by A, and so on. In the case of fighters a propeller-driven aircraft used a single syllable word (the Lavochkin La-11 for example was codenamed *Fang*) while jets had two-syllable codenames (the Lavochkin La-15 was *Fantail*) and this system was used to cover both current and new types. Variants of the same aircraft type received additional letters – for example the initial production Mikoyan MiG-21 was *Fishbed-A* while the heavily upgraded MiG-21*bis* from the 1970s was *Fishbed-L*.

ALEKSEYEV (OKB-21, Gorkii)

I-210 Project for jet fighter first proposed as **I-21** 1946. I-210 to have had German BMW 003 engines but abandoned.

I-211 Jet fighter first flown autumn 1947 (possibly 13.10.47). Development overtaken by versions with more powerful engines. I-211 = I-21 Version 1.

I-211S Swept-wing development of I-211. Design study only, c1947.

I-212 Larger two-seat long-range escort fighter project, spring 1947. Not built.

I-213 Heavier version of I-212 with two British Nene (RD-45) engines. Design study only, c1947.

I-214 Version of I-212 with tail turret removed for radar and three heavy cannon in nose, c1947.

I-215 I-21 version fitted with British Derwent engine (RD-500), 1947. First flown 18.4.48.

I-216 Version of I-211 with two 76mm guns in nose, 1947.

I-217 New development of OKB's twin-engine fighter series, 1947/48. Two versions with either back-swept or forward-swept wings. Not built.

ANTONOV (OKB-153, Novosibirsk – later Kiev)

This was opened in March 1946 as an OKB specialising in the design and development of transport and civilian aeroplanes and was led by O K Antonov. However, during the 1947/48 period and again in 1952/53, this organisation made some brief ventures into fighter design.

Salamandra Light fighter with engine on back of fuselage, spring 1947 onwards.

'M' Masha tailless single-seat fighter with twin RD-10 engines, 1947. Redesigned late 1947 to take single RD-45. Project cancelled when prototype near-complete.

E-153 Scale-model glider of Aircraft 'M' also intended to serve as mock-up for full design, 1948. Cancelled just prior to start of flight development.

Supersonic Fighter Delta wing proposal, 12.52. Preliminary project only.

BAADE
(OKB-1, Dessau, Podberez'ye and later Kimry)

Design team formed with German engineers in Russia at the end WW2. It relocated to Dresden in 1953.

EF 126 Fighter project begun prior to end of war by Junkers company. Restarted by USSR and first gliding flight 5.46. Powered flights followed, but project abandoned 6.48.

EF 137 Jet fighter project, 3.10.46.

BEREZNYAK-ISAYEV
(Engineers working for Bolkhovitinov OKB)

BI Experimental rocket-powered interceptor first flown under own power 15.5.42.

Interceptor Mixed power rocket/jet interceptor proposal by Bereznyak, 1948. Not built.

CHERANOVSKIY
(Small OKB based within MAI Moscow from 1947)

Jet Fighter Flying wing proposal, 1944.

BICh-24 Tailless delta jet fighter proposal, 1948.

BICh-25 Jet fighter proposal with variable-sweep swing wings, 6.48.

BICh-26 Delta-wing jet fighter intended to be capable of supersonic performance, late 1940s.

FLOROV (Engineer working at NII-VVS)

4302 Rocket-powered research aircraft first flown under own power 8.47.

GUDKOV

Gu-VRD Jet fighter project with RD-1 engine, 1942.

KONDRATYEV

A small design group or 'Mini-OKB' run by V V Kondratyev. It is possible that the project described below was drawn while Kondratyev was working with the Yakovlev Design Bureau.

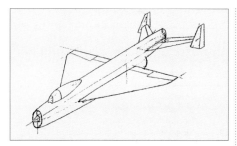

Sketch showing the layout of the Kondratyev supersonic fighter (1953). Russian Aviation Research Trust

Supersonic Fighter Brief study for supersonic delta-winged aircraft, 1953. Swept horizontal tailplane with tip fins, nose intake and two AM-11 engines mounted one above the other with upper jet pipe at end of fuselage, lower terminating ahead of tail. Estimated top speed around 1,700km/h (1,057mph) at height, ceiling 18,000m (59,055ft).

KOSTIKOV (Director of Reaction Engine Research Institute – RNII)

302 Mixed-power fighter first proposed 1940. Glider tests begun 8.43 but ramjets dropped that year, leaving just rocket motor for power. Project cancelled 3.44 before powered flight under way. Also known as **Ko-3**.

LAVOCHKIN (OKB-301, Khimki – was OKB-81, Moscow until 1945)

In 1960, following the death of its chief Semyon Alekseyevich Lavochkin, this OKB became a specialist missile manufacturer. Little or no information is available regarding the gaps in the project number list, but it appears that most of them did not exist as aircraft projects and designs and so were never allocated. Most published sources designate the OKB's project numbers as La-150, etc, but in the first days of the Bureau's jet fighter design these projects were actually known as 'Aircraft 150' (or, more accurately, *Izdeliye* or 'Article 150').

La VRD Twin-boom jet fighter study 2.44. Not built.

'130' Piston fighter prototype first flown 16.6.46. Entered service as **La-9**. ASCC recognition codename *Fritz*.

'138' Experimental version of La-9 fitted with two RD-13 pulsejet engines. Flight-tested spring 1947.

'140' OKB's last piston-engined fighter. Prototype first flew 5.47. Entered service as **La-11**. ASCC recognition codename *Fang*.

'150' OKB's first jet aircraft, a 'lightweight' single-engined fighter. Begun 4.45 and first flown 11.9.46. One example had modified wing as **'150M'**. Second prototype fitted with primitive afterburning to engine as **'150F'**, first flown 25.7.47. '150' no longer active programme by end of year. Production batch of eight (including prototype) unofficially known as **La-13.**

'152' Fighter prototype first flown 5.12.46. No production.

'154' Modification of '152' with Lyul'ka TR-1 engine, late 1946 onwards. Prototype never completed.

'156' Version of '152' fitted with afterburning and other structural changes, 22.11.46 onwards. Initially known as '152D', redesignated 23.12.46, first flown 1.3.47.

'160' (First use of designation). 'Heavy' twin-engined fighter project to run alongside and as insurance to 'lightweight' '150', 4.45. Project shelved 1946.

'160' (Second use of designation). Development of '154'/'156' fitted with swept wing – first Soviet aircraft with wings swept back. First flight 24.6.47. Prototype only which proved to be valuable research aircraft.

'162' Rocket fighter project, 1946. Not built.

'168' To date all OKBs' jet fighter designs actually built had engine underneath forward fuselage with exhaust pipe beneath rear fuselage forward of fin. '168' first to have conventional fuselage with nozzle in tail end of fuselage. Prototype first flown 22.4.48 but rejected in favour of MiG-15.

'174TK' Version of '156' fitted with thinner wing. Flown 1.48 but, despite proving to be an advance over '156', was slower than '160' with swept wing and abandoned.

'174D' Fighter based on '168' but with smaller British Derwent engine instead of the 168's British Nene. Prototype flew 8.48. In modified form (slightly lower tail plus anhedral on wings) entered service in limited numbers as **La-15**. ASCC recognition codename *Fantail*.

'176' Development of '168' with 45°-sweep wing. Prototype flown late 1948 and dived faster than sound 26.12.48. No production.

'190' Transonic fighter prototype, first flown c10.2.51. Intended to carry nose-mounted radar but only eight flights completed before project abandoned.

'200' Two-seat all-weather interceptor prototype first flown 9.9.49. Second prototype fitted with radar. **'200B'** development with greater range and bigger radar flown 3.7.52 but no production. Competitor to Mikoyan I-320 (second), Sukhoi Su-15 (first) and Yakovlev Yak-50.

'250' OKB's biggest interceptor. Initial layout mid-1953 based on '200' but later switched to all-new design. Prototype flown 16.7.56 but bad flying characteristics and a crash forced major redesign. Second modified prototype flew 12.7.57 as **'250A'**. Three more prototypes but project abandoned 7.59.

MIKOYAN (OKB-155, Moscow)

At the time of writing the following lists all known Mikoyan aircraft and projects, but it seems likely that others would have been drawn (particularly after the mid-1950s) which have yet to be found in the archives. However as noted before, a remarkable number of proposals were actually built as prototypes during the 1950s and early 1960s. Any new Mikoyan design received a preliminary designation in the *Izdeliye* category (in the 1940s and 1950s a letter but by the 1960s a number) and then another in the I-series (*Istrebityel* or fighter). These thus form a sequential list to all Mikoyan fighters from World War Two until the MiG-21.

I-250 Piston fighter with additional jet boost system. Also known as *Izdeliye* **N**, first flown 3.3.45, small production batch built as **MiG-13**.

I-260 First jet fighter project, 5.45. Based on German Me 262 with engines under wings, but abandoned and replaced by I-300. Designation *Izdeliye* **F**.

I-270 Rocket interceptor prototype. First example completed flights as glider 12.46. Second made first powered flight 1.47 but damaged soon afterwards and not repaired. Also known as *Izdeliye* **Zh**.

I-300 Fighter with two engines in fuselage, first flown 24.4.46. Designation *Izdeliye* **F** also re-used on this design. Production version entered service as **MiG-9**. ASCC recognition codename *Fargo*.

I-310 Swept-wing fighter project first flown 30.12.46. Also known as *Izdeliye* **S**. Entered service as **MiG-15**. ASCC recognition codename *Fagot*.

I-320 (First use of designation). MiG-9 fitted with British Nene engine, 1947. Prototype abandoned partly complete when realised I-310 far superior.

I-320 (Second use of designation). Radar-equipped fighter also known as *Izdeliye* **R**. Two prototypes built, first flown 16.4.49, but did not enter production. Competitor to La-200, Sukhoi Su-15 and Yakovlev Yak-50.

SP-1 Variant of MiG-15 fitted with Toriy radar in nose radome. First flown 23.4.49. No production.

I-330 Aerodynamically improved version of MiG-15 known as *Izdeliye* **SI**. First flown 14.1.50 and entered service as **MiG-17**. ASCC recognition codename *Fresco*.

SP-2 Variant of MiG-17 fitted with Korshoon radar in nose radome and reheated VK-1F. First flown 3.51. No production.

SN Version of MiG-17 with nose intake replaced by solid nose and side intakes, plus three pivoting 23mm cannon in sides of nose. First flight mid-1953 but showed poor performance, so abandoned.

I-340 MiG-17 prototype rebuilt as test bed for two AM-5A axial engines instead of MiG-17's single centrifugal VK-1A. Flown 12.51. Also known as *Izdeliye* **SM**.

I-350 New design over MiG-17 with Lyul'ka TR-3 engine. Also known as *Izdeliye* **M**. Prototype only, flown 16.6.51.

I-360 Parallel design to *Izdeliye* M fitted with two Mikulin AM-5As and called *Izdeliye* **SM-2**. First flown 24.5.52, two prototypes only.

SM-9 Second I-360 modified and rebuilt as prototype for aircraft that entered service as **MiG-19**. First flown 5.1.54 and OKB's first aircraft to be supersonic on level. ASCC recognition codename *Farmer*. **SM-7** prototype for interceptor **MiG-19P** version flew 24.8.54. Production MiG-19Ps named *Farmer-B*.

SM-12 Series of modified MiG-19s fitted with new engines and improved supersonic inlets. Flight-tested during 1957/58.

I-370 Experimental interceptor with single large VK-7 engine first flown 16.2.55. Also called **I-1**. Fitted with new wings 1956 and became **I-2**.

I-380 'Frontal Fighter' project with single VK-3 engine, mid-1953. Also known as **I-3**. Airframe completed in original form but never flown.

I-410 Experimental interceptor with single VK-3 engine, 1953/54. Also known as **I-3P**, airframe very similar to I-380. Completed but never flown. Modified as I-7U 1956.

Ye-1 Swept-wing proposal made to requirement for new fighter, 1953/54. Planned AM-11 engine delayed. Fitted with AM-9B to become Ye-2.

Ye-50 Development of Ye-1 fitted with auxiliary rocket to compensate for shortfall of turbojet power. Three built, first flown 9.1.56. Known initially as Ye-1A. No production.

Ye-50A Prototype of Ye-50 development with large ventral tank. Never completed and programme abandoned 1957.

Ye-2 Swept-wing proposal for new fighter fitted with AM-9B engine. First flown 14.2.55. Prototype only. **Ye-2A** had AM-11 engine, first flown 17.2.56. Displayed at Tushino 24.6.56 so swept-wing Ye-2A codenamed *Faceplate* by West. MAP expected order for production so allocated **MiG-23** designation to swept-wing version late 1955. Pre-production batch of Ye-2As built.

Ye-4 Delta-wing proposal for new fighter fitted with AM-9B engine. First flown 16.6.55. Prototype only.

Ye-5 Delta-wing proposal for new fighter fitted with AM-11 engine. First flown 9.1.56.

Ye-6 Delta-wing format chosen for development late 1956. This development had Tumanskiy R-11F-300 engine and first flown 20.5.58. Also known in-house as **I-500** and entered service as **MiG-21**. ASCC codename *Fishbed*. Built in many versions and various upgrades still being proposed and tested in 1990s.

Ye-7 Missile and radar-equipped MiG-21 development flown 10.8.59.

'Starfighter' Project combining Ye-1/Ye-6 fuselage with straight flying surfaces similar to American F-104 Starfighter, c1956.

I-420 Modification of I-380 first flown around mid to late summer 1956. Redesignated I-3U (and in some documents I-5). Prototype only.

I-7U Experimental interceptor first flown 22.4.57 with Lyulka AL-7F. Only prototype converted into I-75 1958. **I-7K** derivative with Almaz radar and twin radomes never built.

I-75 I-7U rebuilt with AL-7F engine, first flown in this form 28.4.58. Unfinished I-7K airframe fitted with AL-7F-1 known as **I-75F**. Flight-testing closed 11.5.59. Requirement filled by Sukhoi Su-9.

Ye-8 All-weather development of MiG-21 fitted with nose radar, chin intake, Metskhvarishvili R21F-300 engine, canards and tail. First flown 17.4.62. Two prototypes only because abandoned after engine proved unreliable. Proposed **Ye-8M** upgrade (1962) not built.

Ye-8F Version of Ye-8 with additional lift engines for STOL performance, 1963. Not built.

'7-31' Unbuilt version of MiG-21 with LERX, c1960s. **'7-33'** version of 1976 had RD-33 engine.

Ye-150 Prototype high-speed research aircraft with single Tumanskiy R-15 engine first flown 8.7.60.

Ye-151 Heavy interceptor version of Ye-150 intended to carry the TKB-495 or TKB-539 cannon in tilting mounts, c1959. Fuselage mock-up built.

Ye-152A Interceptor development of Ye-150 fitted with two Tumanskiy R-11F-300 engines. First flown 10.7.59. Codename *Flipper*.

Ye-152 Aircraft intended to combine Ye-150 engine with Ye-152A avionics to form definitive heavy interceptor. First flown 21.4.61. Second prototype converted into Ye-152M.

Ye-152P Development of Ye-152 with more advanced weapon system, R15B-300 engine and canard foreplane, c1961.

Ye-152M Second Ye-152 rebuilt as Ye-152P to serve as prototype, but completed aircraft differed from planned 152P, so redesignated Ye-152M.

Engine still unreliable. Ye-152M abandoned after appearance of MiG-25 and Tu-128.

Ye-155 Programme initiated 1961 covering **Ye-155P** interceptor and **Ye-155R** reconnaissance aircraft. Layouts with twin engines one above the other in rear fuselage, stepped engines or underwing engines rejected for side-by-side position in fuselage. Ye-155R also examined with variable-sweep wing or with lift jets but rejected. Eventually both versions used very similar airframes. Ye-155R prototype first flown 6.3.64, Ye-155P 9.9.64. Entered service as **MiG-25** (codename *Foxbat*). Production also known as '*Izdeliye* 84'.

23-31 Also known as **Ye-7PD** and '*Izdeliye* 92', was example of MiG-21PFM variant rebuilt to house two RD-36-35 lift jets. Flew 16.6.66, codenamed *Fishbed-G*.

MiG-23M Fighter project with side intakes leading to 23-01, 1963. Different versions with square or round intakes, tail of Ye-8 also used initially but design not built in this form. Designation later used by production version of MiG-23.

23-01 Prototype STOL fighter fitted with delta wing and lift jets first flown 3.4.67. Abandoned but displayed publicly 9.7.67 so received codename *Faithless*.

23-11 Prototype fighter with variable-geometry wings first flown 10.6.67. Put into production as **MiG-23**. Codenamed *Flogger*.

MiG-21I Two MiG-21 airframes fitted with wing designed for Tu-144 supersonic airliner for research. Also known as 'Analog' and **21-11**. First flown 18.4.68. Intended to assess wing for civil use but flew well and therefore considered for fighter and ground-attack versions.

MiG-27 Attack aircraft based on MiG-23 fighter. First flown as MiG-23B 20.8.70. Codename *Flogger-F*, etc, sharing the original with MiG-23.

MiG-23A Modified MiG-23 for naval and carrier operations, 1972. Not built.

MiG-23K Further modified MiG-23 for naval and carrier operations, 1977. Not built.

Ye-155M First used for upgraded MiG-25 proposal but designation allocated to more advanced replacement, 1968. Later redesignated **Ye-155MP**. Comprised three initial proposals. Version 'A' based on MiG-25PD, 'B' with VG wings and 'C' with large ogival delta wing and no horizontal tail.

518-21 Fixed-wing interceptor project for Ye-155MP studies, 1968.

518-22 Fixed-wing interceptor project for Ye-155MP studies, 1969.

518-31 Fixed-wing interceptor project for Ye-155MP studies, c1969.

518-55 Fixed-wing interceptor project for Ye-155MP studies with MiG-29-style wing, c1969/70.

Ye-158 Possibly Version 'C' of Ye-155MP.

Ye-155MP Also known as '*Izdeliye* 83', designation now covered layout chosen from above projects, 1972. First prototype flown 16.9.75 and production aircraft labelled **MiG-31**. Codename *Foxhound*. Modified **MiG-31M** flown 21.12.85.

PFI Advanced tactical fighter required to follow MiG-21. First studies 1971, overall project called **MiG-29**. Several configurations examined for both light and heavy fighters. Heavyweight design submitted to 1972 competition and lightweight project two months later. Development of latter

designated '*Izdeliye* 9-12' ordered as prototype – selected ahead of Yak-45I. Heavy MiG-29 design rejected in favour of Sukhoi T-10.

MiG-29 First prototype flown 6.10.77. Codename *Fulcrum*. **MiG-29M** much modified aircraft proposed early 1990s. **MiG-29K** naval carrier version flown 23.6.88. **MiG-29SMT** version of MiG-29 with enlarged dorsal spine to accommodate more fuel, first flown 22.4.98.

33 Single-engine light fighter/attack design with ventral intake, late 1970s. Not built.

5-12 Studies for fifth-generation fighter to meet MFI (multi-role tactical fighter) requirement, 1983. To be twin-engined aircraft. In due course redesignated '*Izdeliye* 1-42'.

4-12 Studies for next-generation fighter to meet LFI (light tactical fighter) I-90 (*Istrebityel*/fighter for 1990s) requirement, c1983. MFI and LFI to share as many components as possible and both received official approval in 1987. LFI to be single-engined but shelved around late 1980s to allow OKB to concentrate on MFI.

1-44 One-off proof-of-concept/technology demonstrator aircraft for 1-42 programme. Almost complete by mid-1994 but not flown until 29.2.00. Programme now abandoned. Codename *Flatpack*.

701 Ultra-long-range interceptor project first conceived in late 1980s to replace MiG-31. Not built. Current status unknown.

LFS Designs for fifth-generation fighter, around mid-1990s onwards. Included single and twin-engined designs.

New Fighter Requirement Full proposals submitted to multi-role fighter requirement 2001. Design from Sukhoi chosen as winner 2002.

MOSKALYOV (OKB-31, Voronezh)

SAM-29 Rocket-powered all-wing interceptor project also known as **RM-1**, 1945. Work ceased at end of war. No detail drawings.

MYASISHCHEV (OKB-23, Moscow)

Just the one design falls within the category of fighter. In fact it originated from Yakovlev and relatively little work was done.

M-33 Designation applied to former Yak-1000M single-seat delta-wing research aircraft with one TRD-5 jet transferred to Myasishchev OKB 3.51. Not built.

OKB-2 (Podberez'ye/Dubna)

This outfit had a relatively short post-war existence and specialised in bombers and rocket-powered aircraft. It was formed in October 1946 and initially led by German designer Hans Roesing. The OKB was closed in 1951.

'346' Supersonic rocket-powered research aircraft derived from DFS 346 under construction in Germany. Flown as glider 1947. First flown under own power 10.50.

'468' Supersonic rocket fighter-interceptor project, 1949. Not built.

'466' Scale model glider of '486' intended to test low-speed performance. Built 1950. Also served as mock-up.

POLIKARPOV (OKB Novosibirsk)

Malyutka Rocket-powered interceptor project 1943. Prototype construction halted 1944 after death of Polikarpov.

SHULEIKOV

Jet Fighter VTOL project with centrifugal jet engine, late 1940s/early 1950s. Not built.

SUKHOI
(OKB-51, Moscow – OKB-134 prior to 1949)

At the time of writing the following lists all known Sukhoi aircraft and projects but, as with Mikoyan, it seems likely that there were others (at least from the mid-1950s onwards) that have not yet been found. The Sukhoi OKB was closed in November 1949 but restored in 1953 after the death of Soviet leader Joseph Stalin, which means that some designations have been re-used. Even without this, the designations used by this OKB appear to be as random and complex to decipher as those from any other aircraft design team in the world. Designations such as 'L' should properly be written as '*Izdeliye* L'. Post-1953, 'S' designations generally related to aircraft having swept wings while 'T' referred to delta-winged designs.

Jet Engine Powered Aircraft Study for design with 'air-breathing jet engine', 10.42. Not built.

I-107 Mixed-powerplant piston/jet fighter first flown 4.45. Later redesignated **Su-5**.

'L' Twin-jet fighter first flown 13.11.46. Official designation **Su-9** allocated later and also redesignated '*Izdeliye* L', but did not enter production.

'LK' Second Su-9 prototype rebuilt with new engines and other changes. First flown 28.5.47 and officially labeled **Su-11**. Single prototype only.

'KD' Projected tactical fighter based on Su-11, summer 1947. Also designated **Su-13**. Prototypes not started.

'TK' All-weather interceptor version of Su-13, 3.48. Not built.

'M' & 'MK' All-weather fighter projects under consideration by end 4.48. No details.

Escort Fighter Proposal fitted with Klimov VK-2 turbojet to escort Tupolev Tu-4 piston-engined long-range bomber, 1948. Little information available but estimated range 5,000km (3,108 miles).

'P' Interceptor prototype first flown 11.1.49. Official designation **Su-15**. Second prototype not completed. Competitor to La-200, Mikoyan I-320 and Yakovlev Yak-50.

'R' Experimental supersonic fighter. Prototype complete by August 1949 but never flown. Official designation **Su-17**. OKB dissolved soon afterwards.

S-1 Project for tactical fighter first proposed 11.53. Design modified before first flown 7.9.55. Entered service as **Su-7**. Appearance of aircraft like North American F-100 as fighter-bomber prompted switch of Su-7 to fighter-bomber role as **Su-7B** (prototype designated **S-22** by OKB). Built in large numbers. ASCC codename *Fitter*.

S-3 Interceptor fighter based on S-1 but with different avionics and weapons, mid-1953. Mock-up completed but project soon abandoned.

T-1 Tactical fighter project, 10.54. Based on S-1 but fitted with delta wing. Design work completed by 12.54 but work stopped 5.55 in favour of T-3 below.

T-3 Interceptor fighter project, 10.54. Prototype first flown 26.4.56. Production plans abandoned for alternative design with different intake. Also known as '*Izdeliye* 81'. ASCC codename *Fishpot-A*.

PT-7 Experimental interceptor development of T-3 with upper and lower intake lip radomes for Almaz radar. Prototype construction begun late 1955 but not flown until end 6.57.

T-5 T-3 prototype converted to take two Tumanskiy R-11F-300 engines. First flown 18.7.58 but no production. Also designated '*Izdeliye* 81-1'.

T-51 Version of T-5 with semi-circular side inlets, late 1950s.

T-43 Proposed version of T-3 fitted with rocket boosters. Rockets never fitted but prototype completed with new nose and wider rear fuselage, variable intake and AL-7F engine. First flown 10.10.57 and production aircraft entered service as **Su-9**. Codename *Fishpot-B*.

PT-8 Further redesign of basic T-3 airframe with different radar and other changes. First flown 21.2.58. Also known as '*Izdeliye* 27'.

T-47 Development of T-3/Su-9 arrangement with longer nose and new radar. First flown 6.1.58 and production aircraft entered service as **Su-11**. Codename *Fishpot-C*.

P-1 Two-seat heavy interceptor with single engine, delta wing, solid nose radome and lateral intakes. Prototype first flown 12.7.59. Later re-graded as experimental aircraft before project closed when work began on T-37 project.

P-2 Twin-engined version of P-1, 1954/55. Developed in parallel with P-1 but cancelled late 1955.

T-37 High-altitude heavy interceptor, 1958. Rival to MiG Ye-150 series. Prototype substantially complete when order made to stop work in 1961. Also known as T-3A and possibility redesignated **T-9M**.

P-37 Alternative version of T-37 with side air intakes, c1958/59. Not built and little information available.

T-49 Prototype produced with alternative nose intake first flown 1.60. Not repaired after in-flight accident 4.60.

T-58 Interceptor project to replace Su-9 and Su-11, late 1960 onwards. Larger radar, side intakes and single AL-7F-2 engine. Construction of two prototypes begun 1960 but halted when Air Force changed its requirements. Also known as **Su-11M**.

T-58 Version of original T-58 proposed late 1960 with two R-11F2-300 engines. Two prototypes built using unfinished single-engined T-58 airframes. Known as T-58D and first flown 30.5.62. Entered service as **Su-15** *Flagon*.

T-59 Interceptor fighter project, 1958. Alternative configuration to T-58 with lateral intakes and based on T-37 but not built. Little information available.

T-60 Project similar to twin-engined T-58 but with rectangular intakes, c1958/59. Not built.

Shkval Private venture VTOL tail-sitter fighter project, 1963.

T-58VD Version of T-58 modified for research into lift engine technology. First flown 6.6.66. Prototype only; codename *Flagon-B*.

T-58TM Major upgrade of Su-15 with new wing leading edge extensions, new engines and updated equipment. First flown 2.71. Entered service as **Su-15TM**.

S-22I & S-32 Variable geometry wing development of Su-7 first flown 2.8.66. Entered service as **Su-17** fighter-bomber, codename *Fitter-C*. Further developed into **Su-20** (for export) and **Su-22**.

T-58PS In-depth upgrade of Su-15 with enhanced interceptor performance, 1972/73. Layout introduced ogival wings and officially designated **Su-19**, but not built. Su-19M had more advanced engines. No official interest received.

PFI Advanced tactical fighter – first studies 1970. Blended wing/body **T10-1** design compared against conventional **T10-2** of 1971. Former chosen as superior. Further configurations examined up to 1975 before order placed for prototype. Selected ahead of heavy MiG-29 and Yak-47.

T-10 Prototype heavy fighter first flown 20.5.77. Flight experience revealed design flawed so layout thoroughly revised.

T-10S T-10 with new wing and other important aerodynamic changes. Prototype (seventh T-10) first flown 20.4.81. Entered service as **Su-27**, codename *Flanker*. Has been continually upgraded.

T-10M Much upgraded version of Su-27 also fitted with canard foreplanes. Su-27 modified as prototype first flown 5.85. First production aircraft, designated **Su-35**, flown 28.6.88.

S-54N & S-56 Proposed fighter-bomber developments of Su-17 fitted with fixed 45°-sweep wing, early 1980s. AL-31F engine and new equipment added but not built because Su-17 itself soon to be retired.

Su-27K Navalised carrier-based version of Su-27 fitted with canard foreplanes. First flown 17.8.87. Official designation **Su-33**.

Su-27IB Fighter-bomber development of Su-27 fighter first flown 13.4.90. In service to be known as **Su-32**; also designated **Su-34**. Nicknamed *Platypus*. Codename *Fullback*.

Su-30 Two-seat multi-role version of Su-27 with equally effective air-to-air or air-to-ground capability. First production aircraft flown 14.4.92.

Su-37 Version of Su-27 fitted with vectored-thrust engines. Prototype first flown 2.4.96.

S-37 Single or two-seat multi-role combat aircraft, c1990. Abandoned around 1992 but development contributed to design of S-37 *Berkoot* forward-swept-wing technology demonstrator. Early published sources called this aircraft Su-37.

S-32 Forward-swept-wing fighter proposal with canards and vectored thrust jet nozzles first designed to MFI multi-role I-90 fighter requirement, 1980s.

S-37 Forward-swept-wing fighter prototype resulting from S-32 studies. First flown 25.9.97. Named Berkoot (Golden Eagle). Redesignated **S-47** in 2000 to prevent confusion with Su-37. Codename *Firkin*.

S-54 Advanced jet trainer, 1990. Design refined into radar-equipped light fighter 1996/97 with training reduced to secondary capability.

S-55 Single-engined thrust-vectoring lightweight fighter project based on S-54 advanced trainer and light combat aircraft, c1996 onwards.

S-56 Believed to be single-engined fighter project offered to India in 11.99. Possibly carrier-based.

LFS Designs for fifth-generation fighter, around mid-1990s onwards.

T-50 Sukhoi proposals to multi-role fighter requirement submitted 2001. Selected as winner spring 2002. Project now known as PAK FA programme.

Soviet Secret Projects: Fighters Since 1945

TsAGI (Central State Aerodynamic and Hydrodynamic Institute, Zhukovskiy)

It is known that at least two supersonic flying boat fighter designs fitted with hydroskis were tunnel tested by TsAGI during 1955, the Models 4221 and 4222. A further hydrofoil design, Model 5202, was tested that year as a supersonic reconnaissance and torpedo aircraft.

TSYBIN (OKB-256)

Ts-1   Rocket-powered research aircraft first flown mid-1947. Also designated in **LL-** series. Aircraft tested with straight and swept wings.

TUPOLEV (OKB-156, Moscow)

Only a few fighter projects were produced by this OKB, which is one of the most famous design teams in the world and responsible for many famous bombers and civil airliners. Those fighters that were designed all fell within the 'heavy' category and, in the case of the Tu-28, drew on the OKB's bomber experience.

'128'   Long-range supersonic interceptor first flown 18.3.61. Initial service designation **Tu-28** but 12.63 became **Tu-128** (ASCC *Fiddler*). Updated **Tu-28A** proposed 1962/63 but not built. Upgraded **Tu-128M** flown 24.9.70.

'136'   Preliminary evaluation of subsonic VTOL fighter, 1963-64. Not accepted.

'138'   Advanced development of Tu-28 with new wing, missiles and avionics, 1962 onwards. Wind tunnel tests showed problems with range and endurance, so further modified. Alternative configurations also examined 1963 but work subsequently replaced by 'Aircraft 148'.

'148'   Long-range interceptor and multi-role strike project to replace Tu-28, 1965. Variable-geometry wing. Tu-148-100 with K-100 missiles abandoned, but mid to late 1960s project revived with K-33 missiles as Tu-148-33. Mock-up built but plans abandoned and requirement filled by MiG-31.

DP-1   Long-range 'raider' fighter-interceptor based on Tu-144D supersonic airliner, 1970s.

DP-2   Long-range 'raider' fighter-interceptor based on Tu-22M bomber, 1970s.

YAKOVLEV (OKB-115, Moscow)

The Yakovlev OKB diversified into many different types of aeroplane but still managed to complete plenty of fighter projects. A number of Yak designations were used twice.

Yak-Jumo   OKB's first jet aircraft. Essentially Yak-3 piston fighter fitted with German Jumo 004B jet. First flown 24.4.46. Entered service as **Yak-15**. ASCC recognition codename *Feather*.

Yak-17   (First use of designation). Development of Yak-15. Prototype prepared for flight but never flown. Did not enter production.

Yak-17   (Second use of designation). Development of Yak-15 with tricycle undercarriage, initially called Yak-15U-RD10 and flown June 1947. Entered service as Yak-17. ASCC recognition codename also *Feather*.

Yak-19   All-new jet fighter with all-through jet pipe first flown 8.1.47. Rival to Lavochkin '156'. Two prototypes only because OKB switched to new Yak-25 with licence-built Derwent engine.

Yak-23   Development of Yak-15/Yak-17 arrangement with British Derwent V engine. First flown 7.47 and eventually ordered into production as insurance for problems with MiG-15. ASCC codename *Flora*.

Yak-25   Developed alongside Yak-23 as alternative with all-through jet pipe. First flew 31.10.47 but straight wing made type obsolete against swept-wing MiG-15. Only three prototypes built.

Yak-27   Single-seat fighter of similar configuration to Yak-25. First proposed 2.47 with nose intake, but in 5.47 revised with wing root intakes similar to later Yak-30 trainer (second use of designation).

Yak-29   Small lightweight single-seat fighter, 7.47.

Yak-30   (First use of designation). Essentially a swept-wing Yak-25. First flew 4.9.48 but MiG-15 favoured, so did not enter production.

Yak-40 & Yak-41   Swept-wing aircraft with wingtip-mounted ramjet engines, 1.48. Yak-40 (and **Yak-40A**) had nose guns.

Yak-41   Experimental development of Yak-40 layout with RATOG, 1.48.

Yak-50   More advanced swept-wing fighter intended to carry radar first flown 15.7.49. Performed well but did not progress beyond prototype stage. Competitor to La-200, Mikoyan I-320 and Sukhoi Su-15.

Yak-60   Light single-seat fighter of Yak-50 configuration, 1948. Not built.

Yak-70   Single-seat fighter project with nose intake, 4.50. Not built.

Yak-M   Single-seat lightweight fighter development of Yak-50, 11.50. Not built.

Yak-U   First use of designation. Version of Yak-M designed 4 to 5.51.

Yak-U   Second use of designation covering project with two AM-5 engines, 1951.

Yak-1000   Research aircraft designed to explore supersonic configurations for fighter powered by TR-5 engine. Taxi tests 3.51 revealed poor handling, so never flown. Officially cancelled 10.51.

'Yak-1000' Fighter   Fighter powered by TR-5 engine and to be capable of high supersonic speeds, 1950. After problems with Yak-1000 aerodynamic test bed, project abandoned 10.51.

Yak-1000M   Proposed modification of Yak-1000 intended to cure design faults, 3.51. Design transferred to Myasishchev as M-33 but not built.

Yak-13   Fighter project which formed direct predecessor to Yak-120 (Yak-25) below, 8.51.

Yak-120   Two-seat patrol subsonic interceptor prototype first flown 19.6.52. Entered service as **Yak-25**. ASCC codename *Flashlight*.

Yak-121   Supersonic development of Yak-25 first flown spring 1956. Pre-production batch built as **Yak-27**. Codename *Flashlight-C*. **Yak-27V** mixed-power high-altitude prototype flown 26.4.57 but no production.

Yak-135   Light tactical fighter project, 1950s (?).

Yak-140   Light single-seat supersonic fighter, 1953. Continuation of development leading from Yak-50. Prototype completed 1954 but never flown. Project cancelled 3.56.

Yak-2AM-11   Two-seat supersonic long-range interceptor derivative of Yak-25 family, 6.54. Cancelled 30.3.55 due to unavailability of engines.

Yak-28P   Interceptor derivative of Yak-129/Yak-28 *Brewer* tactical bomber first flown 1960. Entered service 1963. Codename *Firebar*.

Yak-35MV   Single-engined low-level interceptor project, 1958. Construction planned but project abandoned 1960. Twin-engined fighter-bomber version also planned but cancelled.

Yak-33   Proposed multi-role aircraft, c1961 onwards. Alternative configurations studied and bomber, interceptor and reconnaissance version proposed. None built.

Yak-28-64   Example of Yak-28P converted (re-designed) into Yak-28N with engines in fuselage to compete with Sukhoi Su-15. Believed first flown 5.11.66 but flying qualities poor and not proceeded with.

Yak-36   VTOL research aircraft originally known as **Yak-V**. First hover 9.1.63. Codename *Freehand*. Advanced version with radar also proposed.

Yak-36M   VTOL aircraft with primary role of attack but additional fleet air defence duties. First hover 22.9.70, first conventional flight 2.12.70, entered service as **Yak-38**. Codename *Forger*. Alternative layouts also assessed. Yak-36P designation allocated both to pure air defence version and to more sophisticated supersonic fighter also called Yak-36MF.

Yak-36-70F   Supersonic VTOL fighter project, 1970. Two afterburning lift/cruise engines. Not built.

Yak-39   Proposed development of Yak-38 with bigger wing and more lift capability, 1983. Not built.

Yak-41   Various configurations studied from 1975 for supersonic VTOL fighter with different combinations of lift and cruise engines. Work included development of Yak-45 adapted for VTOL capability, 1979. Yak-41M prototypes ordered and first flown 9.3.87. Export variants possibly to be designated Yak-141 but programme cancelled 11.91. Codename *Freestyle*. Proposed developed version with modified fuselage and wings not built.

Yak-43   Proposed supersonic multi-role V/STOL aircraft for land and ship-borne duties, 1986. Used large main engine plus lift jets. Not built.

Yak-45 & Yak-45I   Twin-engined 'light' fighter projects to PFI requirement, 1972.

Yak-47   Twin-engined 'heavy' fighter project, 1972. Very similar layout to Yak-45 but larger overall. In competition with MiG-29 and Sukhoi T-10 to same requirement, but rejected.

Fighter project   Canard swept-wing single-seat fighter design, c1970s. Canards high on side of box intakes, low-set wing, two engines, missiles on underwing and underfuselage pylons. No further information. Model has '77' code on side of fuselage.

MFI   Multi-role tactical fighter proposal with stealth features, canard, delta wing and twin fins, 1980s. Rejected for Mikoyan 1-42.

Lightweight Fighter   STOVL design proposed to LFS requirement (Lightweight Tactical Aircraft), 1990s. In competition with MiG and Sukhoi designs but rejected.

Select Bibliography

Aircraft of OKB Tupolev: Vladimir G Rigmant; Rusavia, 2001.

Die Flugzeuge des Semjon Alexejew: Helmut F Walther; *Flieger Revue* Issue 1, 1993.

Drops in the Ocean – A Miscellany of Soviet Post-War Jets: Piotr Butowski; *Air Enthusiast No 63,* May/June 1996.

Early Russian Jet Engines – The Nene and Derwent in the Soviet Union and the Evolution of the VK-1: Vladimir Kotelnikov and Tony Buttler; Rolls-Royce Heritage Trust Historical Series No 33, 2003.

Early Soviet Jet Fighters: Yefim Gordon; Red Star Volume 4, Midland Publishing, 2002.

Fluge bis an die Schallmauer: Helmut Walther; *Flugzeug Classic,* October 2003.

I Designed the MiG-15: Mikhail I Gurevich; *Aero Digest,* July 1951.

Illustrated Encyclopaedia of Aircraft from the V M Myasishchev OKB Volume 2, Part 2: A A Bruk, K G Udalov, S G Smirnov, B L Puntas; Aviko, 2001.

MiG: Fifty Years of Secret Aircraft Design: R A Belyakov and J Marmain; Airlife, 1994.

MiG Aircraft since 1937: Bill Gunston and Yefim Gordon; Putnam, 1998.

MiG-21 'Fishbed' – The World's Most Widely Used Supersonic Fighter: Yefim Gordon and Bill Gunston; Midland Publishing, 1996.

MiG-25 'Foxbat', MiG-31 'Foxhound' – Russia's Defensive Front Line: Yefim Gordon; Midland Publishing, 1997.

Mikoyan-Gurevich MiG-15 – The Soviet Union's Long-Lived Korean War Fighter: Yefim Gordon; Midland Publishing, 2001.

Mikoyan-Gurevich MiG-17 – The Soviet Union's Jet Fighter of the Fifties: Yefim Gordon; Midland Publishing, 2002.

Mikoyan-Gurevich MiG-19 – The Soviet Union's First Production Supersonic Fighter: Yefim Gordon; Midland Publishing, 2003.

Mikoyan MiG-23/27 'Flogger' Part 1 – Fighter Versions: Alexander Mladenov; *International Air Power Review Volume 14,* 2004.

Mikoyan MiG-29 Fulcrum – Multi-Role Fighter: Yefim Gordon; Airlife, 1999.

Mikoyan MiG-31 – Yefim Gordon; Midland Publishing, 2005.

OKB MiG: Piotr Butowski with Jay Miller; Midland Publishing, 1991.

OKB Sukhoi: Vladimir Antonov, Yefim Gordon, Nikolai Gordyukov, Vladimir Yakovlev and Vyacheslav Zenkin; Midland Publishing, 1996.

OKB Yakovlev: Yefim Gordon, Dmitriy and Sergey Komissarov; Midland Publishing, 2005.

Russian Aviation and Air Power in the Twentieth Century: ed by Robin Higham, John T Greenwood and Von Hardesty; Frank Cass, 1998.

Russian X-Planes: Alan Dawes; *X-Planes 2,* Key Publishing, 2001.

Russia's Air Power in Crisis: Benjamin S Lambeth; RAND Research Study, 1999.

Samoleti S A Lavochkin: N V Yakubovich; Rusavia, Moscow, 2002.

Soviet Heavy Interceptors: Yefim Gordon; Red Star Volume 19, Midland Publishing, 2004.

Soviet Seaplane Jet Bombers: Thomas Mueller with Jens Baganz; *Aerospace Projects Review,* July/August 2003.

Soviet X-Planes: Yefim Gordon and Bill Gunston; Midland Publishing, 2000.

Su-27 Flanker Story: Andrei Fomin; RA Intervestnik, 2000.

Sukhoi Interceptors: Yefim Gordon; Red Star Volume 16, Midland Publishing, 2004.

Sukhoi S-37 and Mikoyan MFI: Yefim Gordon; Red Star Volume 1, Midland Publishing, 2001.

Sukhoi Su-7/-17/-20/-22 – Soviet Fighter and Fighter-Bomber Family: Yefim Gordon; Midland Publishing, 2004.

Sukhoi Su-15 Flagon: Piotr Butowski; *International Air Power Review Volume 1,* 2001.

Sukhoi Su-27 Flanker: Air Superiority Fighter: Yefim Gordon; Airlife, 1999.

The Osprey Encyclopaedia of Russian Aircraft: 1875-1995: Bill Gunston; Osprey, 1996.

Tupolev Long-Range Supersonic Interceptors – Part II: V Rigmant; *Air Fleet No 17,* June 2000.

Working with Sukhoi: Oleg Samoilovich; Published in Moscow, 1999.

Yakovlev Aircraft since 1924: Bill Gunston and Yefim Gordon; Putnam, 1996.

Yakovlev's VTOL Fighters: John Fricker and Piotr Butowski; Midland Publishing, 1995.

Yakovlev's VTOL Fighters: Anatoli Artemyev; *International Air Power Review Volume 10,* 2003.

Yakovlev Yak-25/-26/-27/-28 – Yakovlev's Tactical Twinjets: Y Gordon; Midland Publishing, 2002.

The following articles, reproduced by the *Bulletin of the Russian Aviation Research Group,* were treated as near primary source material, although most have previously been published in Russian language journals.

Aircraft of P O Sukhoi OKB: N T Gordyukov, V N Zenkin and P V Plunsky.

Experimental 'Shkval': I V Yemel'yanov (*AeroKosmicheskoe Obozreniye 2.03*).

For the PVO Fighter Arm – the Ye-50: Nikolay Yakubovitch (*Kryl'ya Rodini 6.01*).

It Could Have Been the First: Gennadii Serov (*Samolety Mira No 3 3.96*). This article describes the La-200.

MiG-23 – The Long Path towards Perfection: Vladimir Ye Il'yin (*Aviatsiya I Vremya 2.00*).

MiG-27 Story: Vladimir Ye Il'yin (*Aviatsiya i Kosmonavtika 20*).

New Information about Early Projects from P O Sukhoi OKB: V S Proklov.

'Non-Standard' MiG Fighters: Y Gordon (*Aviatsa 1-2*).

On Threshold of Sound Barrier – the '176' by S A Lavochkin Reaching the Speed of Sound: Yuri Smirnov (*Kryl'ya Rodini 3.00*).

Su-9, Su-11 and the Soviet Me 262: Vladimir Prohlov (*Aviatsiya i Kosmonavtika*).

Su-15: V Pavlov (*M-Hobby*).

Su-15 Interceptor: Alexandr Vishnyevsky, V Kulachkin and P Plunsky (*Kryl'ya Rodini 7.99*).

Su-37 – Combination of Manoeuvrability of an Interceptor and Strike Power of a Ground Attack Aircraft: (*AeroHobby*).

Sukhoi Su-37: J Hornat (*Letectvi a Kosmonautika 7.93*).

The Defender of the Sky Frontiers – Su-9 Interceptor: Nikolay Yakubovitch (*Kryl'ya Rodini 7.99*).

The MiG Ahead of its Time – the Little Known Ye-8: Lev Berne (*Kryl'ya Rodini 9.00*).

The Piloted Missile – the T-43, T-47 – Su-11 Modifications: N Yakubovitch (*Kryl'ya Rodini 9.99*).

The 'Suitcase' from a 'Gastronome': Alexey Larionov (*Mir Aviatsii 3.99*). MiG-31 article.

The Supersonic Pipes of the PVO 'All-Union Orchestra': V Pavlov (*Aviatsiya I Vremya 6.98*).

Tupolev Tu-128: Yefim Gordon and Vladimir Rigmant (*Przeglad Konstrukcji Lotniczych 27*).

Finally a detailed and unpublished account of all Tupolev aircraft and project designs, compiled from the OKB archives by Vladimir Rigmant, proved to be a vital source of information for the fighter designs from this famous Design Bureau.

Index